VALUE ANALYSIS
To Improve Productivity

VALUE ANALYSIS
To Improve Productivity

CARLOS FALLON

Manager Value Analysis, RCA Corporate Staff
President, Society of American Value Engineers

WILEY-INTERSCIENCE

a Division of John Wiley & Sons, Inc.
New York • London • Sidney • Toronto

To Maureen, whom I happily love, honor, and cherish; and to Lawrence D. Miles, who has provided me with the means to keep her, feed her, send the children to college, travel, and support my creditors in style.

PREFACE

This book is about value, about the value of a car you want to buy, a workbench you decide to make, or a house you want to sell.

I had bought ships, designed machinery, and manufactured electronic equipment, enhancing value in each case, no better and no worse than anybody else. Then Larry Miles, whom you will meet in the book, opened my eyes: "Poor value," he told his followers, "is a people problem."

The method for solving this problem achieves such smooth interaction among the specialists who contribute value that it is a gold mine! The mining equipment is described in the book. It will be of interest to those industrial managers who must increase gross margins despite higher wages and material costs and to design engineers, buyers, cost accountants, quality specialists, industrial engineers, and those men in Marketing and Finance who have their fingers on the pulse of product value. For such employees the book is meant to provide the miner's pick and miner's lantern, low on fuel, for they will not need it long. Once they master the concept of value and learn how to work rapidly and effectively across departmental lines they usually move up, leaving the miner's tools in the hands of their successors.

The book is also meant as a guide to the value specialist who is ready to move from the evangelical, selling, and training phase to the less spectacular but more durable business of systematically improving the value of his company's principal products.

I have tried to provide a road map for applying basic principles and carrying out necessary tasks. There is a mathematical appendix for those

who want to dig deeper, but you can live without it. Such works as I
have quoted are listed in the references at the end of the book. When
there is more than one book by the same author, I give the date of the
book cited.

When you have been helped by as many people as I have, individual
acknowledgments are impossible. I have been instructed, supported, and
encouraged by my bosses, kept on the ball by my secretaries, protected
by my co-workers, and carried on the shoulders of my value analysts.
On top of it all RCA has encouraged me to write this book by letting
their Publications Services and the staff of the RCA Engineer teach me
to write English. I thank them all.

CARLOS FALLON

Cinnaminson, New Jersey
April 1971

CONTENTS

The Design Engineer. Value Analysis in the Factory. Any Department Manager. The General Manager. The Value Specialist. Who Else Uses Value Analysis.

VALUE ANALYSIS
To Improve Productivity

I

FUNCTION OF VALUE ANALYSIS

If a friend tells you that he analyzes value, you may well answer, "That's interesting, I find myself doing it too," and the conversation will turn to cars, watches, or raincoats. On the other hand, when someone says, "I am in value analysis," the response is usually, "What *is* value analysis?" The two words together imply a special kind of analysis in the same way that *garden flower* is a special kind of flower and *flower garden* is a special kind of garden. Value analysis is a special way to analyze economic values. In the words used by Thomas Henry Huxley about science

. . . its methods differ from those of common sense only as far as the guardsman's cut and thrust differ from the manner in which a savage wields his club.

Like many another discipline, value analysis is simply organized common sense. For this reason, perhaps, some definitions include the words *organized, systematic, purposive,* and *planned,* but the English language has a single word to designate a systematic plan for the orderly application of a body of skills. This word is *method.*

Value analysis is a method that provides the means for industrial disciplines to cross departmental lines in a joint study of a product's function in relation to its cost. The formal definition, which is developed in this chapter, is broader in scope, for value analysis has proved its worth in state and local governments, hospital administration, and major construction projects.

1

We are really discussing the improvement of products and/or services by means of value analysis and/or engineering. Now, you know, those are big mouthfuls! To avoid such jawbreakers I ask the reader to accept throughout the book the word *product* for both products *and* services and the term *value analysis* not only for the parent discipline but also for its offshoots such as value engineering, value assurance, value services, value management, value control, and the truly descriptive value improvement.

WHAT DOES VALUE ANALYSIS DO?

Value analysis improves value by relating the elements of utility to their corresponding elements of cost in order to put the money where it will do the most good. To accomplish this it has to do the following:

1. Find out what the customer really wants the product to do for him.
2. Give the customer more of what he does want and less of what he does not want.
3. Design, make, or buy a product to satisfy the customer's wants in proportion to the relative importance he gives to each of them.
4. Do all this at least cost in resources.

HOW DOES IT WORK?

Value analysis works first by examining the sources of information, the validity of the information itself, and the requirements for additional data; then by defining the function or functions of the product, determining the appropriate cost for providing each function, searching for a better way to perform it or a better way to satisfy the need for it; and finally—like any other scientific method—by evaluating the options, verifying results, and preparing a plan for implementation.

Value analysis, regardless of name, location, or sponsorship, has three universal characteristics: (a) it improves value by studying the function rather than the structure of a product, (b) it deliberately stimulates resourcefulness, ingenuity, and inventiveness, and (c) it begins with a formal information phase. The immediate contribution of good information to product value has led value analysis to develop new techniques for gathering, verifying, appraising, and utilizing information.

WHAT ARE THE GOALS?

Originally product value was analyzed to provide the necessary function at least cost. The Model-T Ford and the first Volkswagen are good examples. Both sold well in market sectors in which least cost was the prime consideration; yet Ford and Volkswagen found greater markets by improving their products within a reasonable cost. In a growing economy least cost is not the customer's major consideration. As his discretionary income rises the customer demands something better every year. Today's value analysis, therefore, is as much concerned with developing new and better ways of satisfying the customer as it is with reducing cost.

✓ Combating the Profit Squeeze   → 10

Wages and the cost of materials are the inexorable costs that must be met day after day in cold, hard cash. Cost reduction must keep pace with these rising costs simply to stay in business. To maintain profit efficiency must be improved and waste reduced at a greater rate than the rise of labor and material costs. If that is not possible, then the product must earn a higher price in the market.

THE RACE BETWEEN PRODUCTIVITY AND RISING COSTS

Why do industrial managers in those nations in which productivity is improving faster than costs are rising nevertheless pay substantial fees to consultants from the United States where productivity is lagging behind the rise in manufacturing costs?

"Because," I was told in Europe, "you have a problem with increasing costs, yes; but not with productivity."

This was a revelation. "We have no problem with productivity?"

My host laughed, serving me smoked salmon and pouring cold aquavit. "Your productivity is nearly perfect. That is why it is hard to improve. We are winning the race with costs because we still have room to improve our productivity. That is why we invite North American consultants to help us bring our productivity up to the near-perfect level of yours."

I hated to tell him that I knew a dozen good men who had more recent experience with finding better ways of making a given product. "You understand," I asked uneasily, "that I am a value analyst?"

"Exactly!" he plopped a gigantic slice of smoked salmon on my plate. "This is a different kind of salmon," he explained, "the other one was smoked just like he comes out of the sea. This one is jellied before he is smoked."

Wondering whether I would be smoked or jellied or both, I picked uneasily at the fish.

"You see," my host was explaining, "our factory-floor productivity will soon be approaching the level of yours, leaving little room for improvement; yet our workers will demand and deserve higher wages. When we can no longer improve the ways of making the product, we want to be ready to work on the product itself, using *vardeanalys*."

Now I was eating the fish with enthusiasm. I knew that most customers are not getting what they really want. Give it to them and the product earns a better price. The customers have the money to pay for it, too, out of the wage increases that industry has been paying *them*. Productivity has increased because the workers are turning out a better product for the same cost. The results of their labor and their imagination are worth more than the results of their labor alone.

As my host explained, "A higher price for the same product is inflationary, but a higher price for a better product is not. Our workers want higher wages because they want a richer and fuller life. So we give them better products to make their lives richer and fuller and they give us more money for the better products."

When you can no longer improve the way to make the same old product, you can still make a better product. The question then changes from "What is a better way to make the product?" to "How can we make it a better product?" To answer that question the whole industrial team has to get into the act. That's what value analysis is all about: better value and suitable levels of inventory, price, and quality; for as the U. S. Department of Commerce booklet *Profits and the American Economy* so ably puts it,

. . . producing too much or too little, or at a price that is too high or too low, or at a level of quality uncalled for or inadequate, can wipe out profits.

The function of value analysis, however, is broader than we anticipated. We were looking for, and achieving, a better relationship between what the customer has to pay and what he gets for his money; but what the customer gets for his money is not provided by any single industrial discipline. Poor value is often caused by poor interaction among the contributing disciplines—not poor cooperation—but poor information and poor communication. Forestalling problems of information and preventing failures in communication are primary tasks of value analysis.

When a declining share of the market reveals that the customer is unhappy with the performance, quality, style, or price of a product, there is a crisis in the company; everybody gets into the act, and the joint effort of diverse specialists produces a better product—all as the result of an emergency.

Why not on purpose?

Because an industrial organization, although hopefully called a team, lacks some of the characteristics of a team. Like a team, it combines diverse skills with a pattern for their interaction; unlike a team, the interaction is seldom left to the players who are physically handling the ball.

Improving the Industrial Team

Traditionally, Engineering designs a product at the request, direction, or suggestion of Marketing and Styling; Finance provides the resources and rides herd on them; Purchasing buys materials and some specialized design services; and Manufacturing makes the product.

But here is the rub. Engineering and Manufacturing are expected to work mainly through the physical sciences, whereas buying, selling, styling, and financing are part of the social sciences, or so the universities have taught us. Faithful to the traditional classification of knowledge, we march our separate ways, forgetting that classification does little to the real relationships among the elements classified.

No matter how we separate Engineering and Manufacturing from Finance, Marketing, Styling, and Purchasing in the organization, the value of our products depends on all of them together, on their effective interaction as much as on their departmental contribution. Two requisites for working well together among departments are timely information and simple, straightforward communication.

In order to provide a mechanism for face-to-face communication among the players who actually handle the ball the key specialists *at the working level* are brought together in task groups which can truly be called teams. Such teams exchange and digest in a few hours the information that normally takes weeks to circulate in written, drawn, or coded form.

Duplication of effort? Not at all. Value analysis does not pretend to do or to correct the work of designers, buyers, production men, and cost estimators. These specialists apply their skills to their usual tasks, but in a value task group they apply them jointly and often concurrently, thus saving precious time.

Is the task group approach the best way to improve product value?

Not always. I have reporting functionally to me, among others, three thoroughly experienced and successful analysts, none of whom uses the task-group approach. One of them has no choice; his specialists are scattered over mountaintops, desert stations, and the islands of the sea. Another one combines, within himself, electronic, mechanical, and production engineering skills as well as experience in cost accounting and administration. He circulates among the specialists until they all understand the effect of their information on the work of the others. When all the facts jibe, he does his own value analysis and submits recommendations to the responsible decision makers. The third one is a loner, extremely independent, extremely studious, and extremely competent in all aspects of the product made in his division. His recommendations stand up on their merit alone.

These men are exceptional types, hard to come by. Since we cannot expect such unusual skills from everybody, we generally use the task group approach in which the special skills remain in the hands of the regular specialists, the value analysts simply providing a mechanism for timely interaction and for guidance in applying the method.

Both the individual approach and that of the task group work well in small plants as well as large ones. It is well to point out that a decentralized corporation such as RCA has many small plants that resemble small business operations.

COST REDUCTION AND VALUE ANALYSIS

Larry Miles originally applied his method to the elimination of unnecessary costs. His approach to product value turned out to have greater capabilities, but when the method is applied to the sole objective of reducing cost *it is* cost reduction! . . . And what is wrong with that?

We have all felt the impact of rising costs. An increase in costs, without a corresponding increase in productivity, immediately weakens a company's cash position. Conversely, an effective cost reduction action reduces the outflow of cash. When the inflow continues, there is an increase in ready money. One of the accomplishments of value analysis is cost reduction, but cost reduction is so important that there are many activities dedicated exclusively to it. Why? The answer is cash.

Importance of Ready Cash

Being an instrument for quick action, cash exerts a potent leverage on profit. A few dollars in the form of immediate ready money are often more useful than many dollars already committed; and there is no quicker source of ready money than the money already in your pocket. This is the cash that cost reduction makes available for instant use. Such cash provides a means of grasping opportunity—for taking the offensive.

Good Costs and Bad Costs

Not all that glitters is gold, nor are all costs bad. Neither is all cost reduction equally good. Reducing all costs by the same percentage is like punishing all crimes with the same sentence. Both instances avoid thinking and controversy.

A judge can say, "Today I am on a crime-reduction drive. I am going to hang everybody! Don't give me any guff. You are all being treated alike."

While the gallows are taking care of murderers and parking violators, the judge's brother, in a factory on the outskirts of town, announces: "I am on a cost reduction drive. All costs will be reduced, the costs that sell the product as well as the cost of parts no longer needed."

There are costs and costs, and we have to study them in relation to the competitive position of our products; otherwise we cannot protect profitable investment money from the axe meant for waste.

THE NATURE OF COSTS

Think of the business as a black box in systems engineering. The outputs are profitable products and the inputs are costs that represent resources consumed. As significant as their level is the nature and the number of these inputs. We have to consider all costs—everything that must be committed or used up to produce the output we want. Usually quantified in units of currency, these costs represent the use of

> TIME
> SPACE
> MATTER
> ENERGY
> HUMAN EFFORT

Time is one of the most significant cost factors; yet its nature is little understood because the word has many meanings. Let's consider these meanings:

Time in the sense of duration exists in two forms. The duration of an effort can be divided by the number of people or machines exerting the effort—man-hours or machine-hours—a factor you can manipulate. Necessary duration, such as the time it takes sheep to have lambs, is hard to change.

Time in the sense of elapsed time is the relentless, inexorable stream that flows steadily until we meet or miss that delivery date. It is inflexible, uncontrollable, and exactly metered out for everyone alike. How much is done by two contenders within the same elapsed time is a factor as deciding in industry as it is in warfare.

The relation of time to causation is noted in the Bible. "For everything there is a season, and a time for every matter under heaven . . . a time to plant and a time to pluck up what is planted." You have to plant first because planting is the cause that engenders the harvest. Time, in the sense of causation, brings out the importance of proper sequencing. Techniques to optimize sequencing, of course, are critical path analysis, Pert, and Pert Cost.

Time in the sense of opportunity is sometimes related to causation, as in the quotation cited above, but it is also related to an inviting situation or condition. The saying, "Strike while the iron is hot," is as good a description of timeliness as I know. Sound information is the basic tool for reducing cost by better timing.

Space

The cost factor here is the cost of volume. Volume governs the number of bulky items that can be carried in a ship, a railway car, or a plane. The bigger they are, the fewer you can stow on board.

Matter

The cost factor associated with matter is mass, either as undesirable inertia or as weight. The greater the mass, the harder it is to start things moving and the harder it is to stop them. Mass determines weight, and weight is the limiting factor in shipping heavy items.

Energy

Cost reduction by the elimination of wasted energy usually yields savings in weight and volume as well as in dollars, and the energy gained can often be used to improve performance.

Human Effort

The great advances achieved by industrial engineering have dramatically reduced the labor content of our products. Two recent industrial engineering programs of reducing the cost of human effort are work simplification and error-cause identification. The latter, associated as it is with pride in craftsmanship and labor-management teamwork, is a distinct step forward in industrial relations.

But there is a kind of human effort which has not received the attention it deserves and that is the human effort of the knowledge workers, the people who decide on the what and when of a company's products. These people need not be senior executives. Many employees at various levels in the modern industrial organization work at getting the right things done as distinguished from simply doing things right. Peter Drucker's book, *The Effective Executive,** is a good guide for improving the performance of knowledge workers. Here is a paragraph from the first chapter:

There are few things less pleasing to the Lord, and less productive, than an engineering department that rapidly turns out beautiful blueprints for the wrong product. Working on the *right* things is what makes knowledge work effective.

PUTTING A DOLLAR SIGN ON COSTS

We cannot put a number on the value of an engineering decision or a financial judgment, but the use of such resources as time, space, matter, energy, and even routine human effort can often be quantified as follows:

TIME:	dollars per month of elapsed time
SPACE:	dollars per cubic foot
MATTER:	dollars per pound of weight
ENERGY:	dollars per kilowatt-hour
HUMAN EFFORT:	dollars per man-hour

INTERACTION OF COSTS

Any line function can improve its contribution to profit by establishing an alliance with the people in Finance in order to determine the interac-

* From *The Effective Executive* by Peter Drucker. Copyright 1967 by McGraw-Hill. Used with permission of McGraw-Hill Book Company.

tion of significant costs. This plan calls for the study of such measures as overhead rate, materials-handling expense, cost per purchase order, and cost of an engineering change. These necessary costs can then be optimized rather than arbitrarily minimized.

By identifying and capturing resources that are going down the drain cost reduction converts useless into usable wealth. When such wealth takes the form of cash, which no longer has to be spent, the impact on operations can be felt right away. The unsung cost reduction specialist suddenly becomes a magician—a creator of resources! He creates wealth in the same way that miners or fishermen create wealth. They do not *make* the minerals or the fish. They find and capture these elusive resources, thus making them available for use.

Importance of Cost Reduction

Everything I have said about cost reduction as one of the aims of value analysis is true about plain cost reduction done under its own name. Value analysis makes it easier to reduce costs and easier to increase costs when it is more profitable to do so, but when the constraints are "Make it like the drawing! Buy from the list of materials! There's no time for changes!" then the cost reduction specialist can do a far better job than the value analyst. The latter would seek relief from the constraints in order to do his thing, but such constraits are often true and necessary. In that case the value analyst throws the ball to the man who is nearest the basket.

But We Have to Work on the Goodies Too

As I mentioned earlier, I was in value engineering before I went to corporate staff. One of the many lessons I learned then under a division vice president who suffered as my superior was that the value disciplines must go beyond cost reduction.

"You value engineers," he told a group of us, "spend half your time saying that value engineering is not cost reduction and the other half saying that if it does not reduce cost it is not value engineering!"

He then gave us two assignments: the first to learn something about value and the second to discover the difference between value engineering and cost reduction.

But then he added, "Maybe you *do* know the difference. What *is* the difference between VE and cost reduction?"

We tried to answer his question with another question: "In cost reduction do you define the function of a product, assign an appropriate value to that function, and then carry out a planned creative effort to provide the necessary function at least cost?"

It is not easy to confuse vice presidents by answering their questions with other questions. He asked, "Why should we?"

"To reduce cost, of course," we all answered in unison.

And we were trapped!

"If the purpose of the technique is to reduce cost," he purred in a tone of voice reserved for the very young or the very old, "then it is cost reduction—" the tone changed—"AND NOTHING BUT COST REDUCTION!"

While we picked up the clock and calendar, which had fallen from the desk as he emphasized his point, the boss drew a deep breath. "If I may use your own terminology," he continued in a calmer voice, "the function of value engineering can be expressed in two words, a verb and a noun—"

Two value engineers from the downtown plant tiptoed in.

"You are just in time," the boss told them. "We were about to define the function of value engineering in a verb and a noun." With disarming gentleness, he added, "Can you do that for us?"

"Reduce cost," they answered together.

Half of us looked at the ceiling, the other half at the floor.

During our prolonged silence the boss had put through a call to the manager of cost reduction. "See what training slots you can open up for people who are skilled in cost reduction but not altogether committed to it."

Gazing at the landscaped grounds through his ample plate-glass window, he sighed. "A tough part of the job—planning demotions and retraining, trying to get the right men into the right slots—"

A cloud obscured the sun; the room grew dark.

"Do you mean, sir," someone asked in a quavering voice, "that there will be no slots for value engineers?"

"For value engineers?"

"Yes, sir."

"Of course," was the answer. "I need replacements for you boys, but don't worry about my problems—" he motioned to the coffee and doughnuts which had just arrived—"Relax, have some coffee. I'll make out all right. It will take me at least a month to place you in training slots. During that time Personnel can find me people who know something about value."

It was a hint. Nobody touched the coffee.

"Oh, come, you are looking at me as though I was crucifying you." The boss began arranging our personnel folders on his desk. "But it takes considerable merit to deserve crucifixion." He looked at us with a hint of fire in his eyes. "I'll give you a chance to rise to the challenge, either crucifixion or glory!"

The day brightened and we began drinking coffee.

"You know, mine is not an easy job," the boss confided. "I watch you gathering all sorts of benefits, through value engineering, but out of the rich catch in your nets you keep only cost reduction. Everything else you throw back into the sea. If you can find out what, besides cost reduction, value engineering can do for an industrial plant, *and if you do it*, you can stay *and* advance in value engineering."

We took the hint and went to work, wringing the full potential out of the discipline, regardless of its name.

Summary of Results

Value analysis accomplishes a modest increase in gross margins by integrating product improvement with cost reduction. In this way better materials can often be bought for the same or less cost, and products that are manufactured to a given cost can *earn* a higher price in the market place.

VALUE ANALYSIS AND VALUE ENGINEERING

Value analysis is the parent discipline, founded by Lawrence D. Miles. "Value engineering is value analysis when applied in the engineering sphere."—Miles.

> *That which we call a rose,*
> *By any other name would smell as sweet.*

> *Romeo and Juliet*, Act II, Scene ii.

Upstream, Downstream, or Both?

Because many people believe that purchasing agents and procurement officers buy only existing products, value analysis in procurement has been described as "second-look value analysis." Much original design engineering is *purchased* by procurement officers and purchasing agents who must organize first-look value programs in the vendor's plants. Conversely, many engineering departments, turning out near-perfect value engineered designs, must ride herd on their products and take a *second* look whenever customer needs change or new materials appear on the horizon.

The truth is that value analysis, by whatever name it is called, must take both first *and* second looks. A manufacturer must analyze the value of his product whenever and wherever it is profitable to do so. There is no place for the either/or mentality in modern business. This is not a matter of definition but of competition.

The Armed Services Procurement Regulation, the Department of Defense Directorate of Value Engineering, and the Society of American Value Engineers all consider the terms *value analysis* and *value engineering* synonymous.

The Department of Defense Handbook 5010.8H (1968), page 1, sums up the label situation very neatly:

> Value Engineering (VE) is the term used in this handbook and by the DoD in its contracts. Others may refer to their value improvement efforts by such terms as Value Analysis, Value Control, or Value Management. There may be some subtle differences between these other programs and VE, but the basic objectives and philosophy appear to be the same for all.

I have chosen the term *value analysis*, first because it appears on my paycheck and second because it does not tie the analysis of value to any one discipline.

I see no reason why the reader should not call the discipline by the term that appears on *his* paycheck or whatever seems more logical to him. But no matter what he calls it, sooner or later he will be asked for a definition.

The first definition of value analysis, and still a classic, is from Miles (1961), page 1.

> Value analysis is a philosophy implemented by the use of a specific set of techniques, a body of knowledge, and a group of learned skills. It is an organized creative approach which has for its purpose the efficient identification of unnecessary cost, i.e., cost which provides neither quality nor use nor life nor appearance nor customer features.

In the American Management Association book, *Value Analysis/Value Engineering,* Falcon (1964), page 10, I wrote:

> Value analysis/value engineering is a functionally oriented scientific method for improving product value by relating the elements of product worth to their corresponding elements of product cost in order to accomplish the required function at least cost in resources.

I used *product worth* here as a monetary measure of utility, that is, of the usefulness, esteem, or both, that makes the customer want a product in a given situation. But who knows what the customer wants in a given situation?

Marketing, of course.

Who creates the concepts of usefulness and esteem to satisfy the customer's wants? . . . Engineering and Styling.

Who finds and procures the necessary materials and outside services? . . . Purchasing.

Who brings the product concept into reality? . . . Manufacturing.

Who provides units of measure to compare the cost and results of these activities? . . . Finance.

Obviously we cannot get a complete picture of product value through the eyes of a single discipline—not even value analysis.

So . . . ?

So value analysis is a multidisciplinary method—no! not multidisciplinary. Such a mouthful would kill the definition. *Multidiscipline* is a shorter word that performs the same function; but adding *multidiscipline* to my detailed definition would make it too cumbersome, so I reworded the whole thing to include multidiscipline in a one-mouthful definition. Here it is:

> *Value analysis is a multidiscipline method for enhancing product value by improving the relationship of worth to cost through the study of function.*

The word function is really the kernel of value-analysis terminology. Later on we shall see how Larry Miles, when faced with a routine problem of substitution, asked the question "What does it do?" and had the insight to build a whole new method of investigation centered on *the function* of the product. Miles pinpointed the dynamic aspect of value and proved, by the success of his method, that what counts in economics is what the product *does*.

This concept frees the mind from structural limitations. It opens the door to variation and innovation, directing thought to the dynamic task of customer satisfaction. The results, though surprisingly profitable, are often obvious decisions which had not been made before because somebody thought that *somebody else* could not comply.

Someone, not in Marketing or Applications Engineering, says, "The customer won't like it."

Someone, not in Manufacturing, says, "It can't be made that way."

Someone, not in Purchasing, says, "It can't be procured."

So nothing is done.

It *could* have been done. It could have been done *easily*. *Anybody* could have thought of the idea. It was the *obvious* thing to do!

But . . .

It was not done!

Dynamic inertia? People working at *not doing* what has to be done? Resisting change, just to resist?

Not so!

I am talking about good people, employees who are dedicated, daring, and imaginative. Why did they not do the obvious?

Was it a matter of willingness, courage, or skill? No. It was a matter of *information and communication*. The very lack of data that would help him distinguish a hole in the ground from a more intimate orifice prompts many an old-fashioned industrial manager to hide behind the excuse, "Our business is different."

Of course it is different. This very uniqueness of many a modern enterprise calls for better methods to relate the "different" business to the not so different goals of profit, growth, and return on investment.

The more exotic and advanced a business is, the greater the need for rapid interchange of information among its industrial departments. Not only are "obvious" decisions delayed by the tortuous travel of technical information but profitable actions are neglected and losing actions continued because cost data does not reach the technical people in time.

During the last 20 years value analysis has developed a body of industrial skills which helps to integrate the efforts of the businessmen, artists, and technical men who are scattered throughout the modern plant. By providing a mechanism for face-to-face communication at the working level value task groups interpret technical information in business terms and business information in technical terms.

With respect to the usefulness of value analysis in identifying and solving problems of information, communication, and inertia, Miles (1968) puts it this way:

Now back to the question of what it is. Value analysis is a superior problem-solving system.

HISTORICAL SKETCH

Value analysis is the indirect result of the crisis in resource distribution in the late 1930's. Short on material resources, the Have-Not-Nations—Germany, Japan, and Italy—went into World War II with the idea of using their highly militarized human resources to take what they needed.

Although the Allies had reasonably good fighting forces, they had a strong preference for keeping their people alive. They chose, therefore, to conserve their human resources by overwhelming the Axis Powers with material resources. In the United States the word went out from Washington, "The best for our boys! The most of the best of everything for the war effort!"

The United States was embarked on a deliberate policy of lavish use of resources—a very sensible policy *at the time*. By 1947 we began feeling the backlash of that policy. To begin with, there were not so many material resources around as we thought. Shortages began to appear everywhere. Certain materials simply were not available; others had skyrocketed in costs.

At the GE Plant in Schenectady Larry Miles, the man who founded the discipline of value analysis, was faced with the technical problem of successful substitution and with the economic problem of high cost. Now when you attempt to substitute, you are looking for *something else,* and you are immediately freed from all the restrictions associated with the original product or service. All you have left is the function. You are now concerned only with what the product or service *does* and with *what else* will do the job.

By concentrating his efforts on fully understanding *the function* of a product or service Larry Miles laid the foundation stone for value analysis. He then went on to develop means of "providing the performance the customers want and for doing so at an appropriate cost."

In his efforts to improve the interaction between the people responsible for the physical and the economic aspects of a product Miles discovered that poor value is "a people problem." So he developed an arrangement of techniques—some old, some modified, some new—for the early and fruitful interchange of information among the diverse disciplines that contribute to product value.

The system worked surprisingly well. It revealed that just knowing what things cost could save 5%; improvements in the choice of materials and methods, another 10%, but finding a better way to do what the product was supposed to do in the first place could save 30% and more. From that moment on Miles concentrated on the function and his motto became *define the function; evaluate the function.*

An engineer interprets the word *function* in terms of performance. This approach put value analysis into preliminary design and specification writing.

In 1954 the U. S. Navy Bureau of Ships, under Rear Admiral Legget and his assistant Rear Admiral Mandelkorn, set up a formal Navy value program with guidance and training from Miles himself and his GE team. The Bureau of Ships, primarily concerned with engineering, adopted the discipline under a new name—*value engineering*. In this way it could assign engineers to analyze value without changing their engineering titles. There were civil service slots for engineers but none for "value analysts." We will see later how the term *value analysis* itself originated.

By 1956 all 11 naval shipyards had active value-engineering organizations. Private industry, in the meantime, had been adopting the discipline. In 1952 the RCA Engineering Products Division appointed an administrator of value analysis, and the corporation has had an active value program ever since.

The Army Ordnance Corps Watervliet Arsenal, because of its proximity to General Electric, couldn't help hearing about the savings. They asked for a little neighborly guidance.

With the help of both BuShips and GE, Army Ordnance set up a value-analysis organization in which six arsenals, the Ordnance Ammunition Command, the Army Rocket and Guided Missile Agency, and the Army Ballistic Missile Agency participated. All these value-analysis activities are now part of the Department of Defense value-engineering effort.

Not only did the services unite in their efforts to get better defense for the dollar but defense contractors followed suit, freely sharing value-analysis information with one another. Though state governments have picked up the method to get better use out of their resources, the most significant growth has been in the private sector—from automobiles to motorcycles and from earth-moving equipment to sewing machines, not to mention hospitals, oil companies, and telephone companies. Private industry in Canada has kept right abreast of value-analysis progress in the United States.

Overseas, Sweden and Japan have been the most active in value analysis, with Norway, Denmark, England, Scotland, Ireland, and West Germany following closely. The growth of value analysis "Down Under" has been characteristic of the dynamism of Australia and New Zealand.

The Society of American Value Engineers (SAVE), currently chartered as a national society under the laws of the State of Georgia, was incorporated on October 22, 1959, in the District of Columbia.

Sister societies are the Scottish Association of Value Engineers (also SAVE), the Society of Japanese Value Engineers, the Canadian Society for Value Technology, the Scandinavian Society of Value Analysis (SCANVAVE), and the English Value Engineering Association.

ORIGIN OF THE TERM VALUE ANALYSIS

Accepting John Stuart Mill's solution of the paradox of value, most economists agree on the two essentials of exchange value: "The thing must conduce to some purpose, satisfy some desire" (our function), and "there must be some difficulty in its attainment" (our cost).

The first of the two essentials, the one which would "conduce to some purpose, satisfy some desire," has been considered beyond the scope of measurement—unanalyzable! Russell Ackoff said in his extraordinarily useful book, *Scientific Method* (1962):

The analysis of the "unanalyzable" and the conception of the "inconceivable" have in the past constituted some of the most important spurts in the progress of science.

We owe such an advance to Larry Miles. He not only tackled the unanalyzable but he showed us how to analyze it, how to identify, classify, and evaluate the function, how to pinpoint its most important aspect in a verb and a noun, how to select measurable nouns, and how to combine physical measurements and business measurements to yield an economic specification.

To understand the insight that led to the development of value analysis we should note that Miles has degrees in education *and* in engineering. His own training encompasses the social as well as the physical sciences, hence his early efforts at emulsifying the oil and water of these two arbitrarily separated fields.

Buying and selling is a *social* activity and Miles was a purchasing agent working for Harry Erlicher, Vice President of Purchases at GE. Both naturally understood the "people problems" that led to poor value. Now, engineering deals with the *physical* sciences, but engineers are *people,* so Erlicher and Miles set up a meeting with Harry Winne, Vice President of Engineering. They found an engineer in the fullest sense of the word, a man who realized that in engineering manipulating the sources of power and the properties of matter has only one purpose—the service of man. He had tailored his professional development to accomplish this purpose.

Harry Winne saw the importance of the method at once. "What are you going to call it?" he asked.

Miles had not thought of a name for his new approach, but Winne was enthusiastic. "You are dealing with *value,*" he said. "Value has a performance part and a cost part."

There was a pause. Then Harry Winne asked, "Why don't you call it *value analysis?*"

Miles, who had already taken the step from fact to value when he asked the question, "What *should* it cost," grinned in appreciation of Winne's support. "Value analysis it is!" he announced happily.

SPECIFIC TASKS OF VALUE ANALYSIS ✓

Application of the principles of value analysis follows the job plan outlined in Chapter 4 and developed throughout the rest of the book. The specific tasks that make up the job plan, however, must be introduced now in order to give the reader a better grasp of Chapters 2 and 3.

Table 1-1   Sequence of Value-Analysis Tasks

1. Preparation	2. Investigation
3. Analysis	4. Innovation
5. Simplification	6. Evaluation
7. Comparison	8. Choice
9. Presentation	10. Implementation

PREPARATION. Preparation starts when certain products cry out for analysis because they did not go over so well as they should, competition has hurt them, or they are not yielding enough gross margin. In this case the task starts with the ailing product and its nurse. Preparation for a value-analysis workshop may also start with a manager who prefers to be the one that the competition worries about, the manager who takes his lead products, and pushes them two jumps ahead. Selection of products and participants and other preparations for a value workshop are decribed in detail in Chapter 10.

INVESTIGATION. The task of investigation involves all aspects of collecting, screening, and appraising current information on the product to be analyzed and on finding out what the customer really wants today. It is described in Chapter 5.

ANALYSIS. Analysis involves identifying the functions of the product and assigning an appropriate cost to each to concentrate effort where it will do the most good. This process is described in Chapter 6.

INNOVATION AND SIMPLIFICATION. These tasks, covered in Chapter 7, represent the development of better functions to satisfy the customer or better ways to perform them. Simplification is a necessary step in refining new designs and even design changes.

EVALUATION, COMPARISON, AND CHOICE. These steps constitute the decision process within the value task group. Chapter 8 presents the COMBINEX method for comparing and selecting options and for explaining the selection criteria to higher management.

IMPLEMENTATION. Everybody urges implementation, and we all agree that plans and decisions are *Love's Labor Lost* until they are implemented. There is mighty little information available on implementation, yet I have to lecture on it. I tell the audience what I know about implementation; then, in the question period they tell me what *they* know. Since the lecture victims outnumber the lecturer, I get more out of it than they do. I try out their recommendations—using live projects; I present the results to other groups of thoroughly experienced industrial managers, and when it looks like we all have put together something that really works I try it out a few more times. Then I write it up as Chapter 9 of this book.

ABSTRACTING GENERAL PRINCIPLES

The thought that certain basic principles must underlie the successful performance of value-analysis tasks was farthest from my mind when I was called to the office of my supervisor's boss in 1960. Without preamble the big boss said, "It looks like there is something behind these value disciplines."

I nodded politely.

"They work," he continued.

"Yes, sir!"

"That's not enough," he snapped.

The expression on my face must have said, "What's not enough?"

The big boss explained, "Doing something right, without knowing why." A conspiratorial gleam brightened his eyes. Motioning me to a chair besides his desk, he directed, "Find out *why* these value disciplines work. Pick out your best examples, compare them, and abstract the general principles behind them."

A tall order.

It took me some 10 years to comply. The road signs which constitute the principles of value analysis originate in a variety of disciplines: economics, strategy, management, and psychology.

A principle is a rule of conduct or guide to action. The principles of value analysis are all implicit in the work of Larry Miles, particularly in his most recent work. All I have done is to identify them and describe their application in the light of my own experience.

Usefulness of the Principles

Had we, in value analysis, understood and applied certain principles from the beginning, our specialty would not have aroused the hostility accorded a brash youth who bursts into the halls of his elders to teach them a new way of doing things.

To say that the good old-fashioned way is no longer good enough is beside the point. Seldom can you get people to improve by putting them on the defensive, by telling them that you know better, or by characterizing their objections as "road blocks in the path of progress."

Cold Hard Facts Can Hurt

The principles of usefulness, limited resources, economy, adequate information, and effective communication, because of their immediate impact on the cost-profit squeeze, rode rough-shod over many an objection. The triumphant value analyst, with his task group serving as a chorus, proved that money could be saved where others said it couldn't. He showed dramatically how wrong everybody was. Finally, he achieved complete unity of purpose among department heads; they all wanted to get rid of him. After the farewell luncheon members of his task groups were left stranded, wondering why nobody loved them, when they had done everything right. Larry Miles, who is at this writing more active and perceptive than ever, sums up the situation this way, "Embarrassment! They had embarrassed the people responsible."

Nothing Wrong With People Acting Like People

Additional principles had to be invoked: direction, responsibility, interacting skills, direct motivation, and objectivity. Application of these principles avoids disrupting the lines of command, avoids encroaching on the territory of others, engenders respect for the man responsible, and shows us how to offer solid information instead of exhortation. Moreover, understanding why people act like people shows us how to put the fuel of useful learning and personal accomplishment under the boiler of human aspirations, keeping out of value analysis the cultist who must remake people instead of learning to do the job with people as they are.

Then the overwhelming evidence of actual consumer behavior revealed that constant product improvement is at least as urgent an industrial necessity as cost reduction.

In the Free World People Buy What They Want

More principles! consumer sovereignty, balance and proportion, selectivity, and concentration all aimed at giving the customer a better product at a fair profit to the manufacturer.

Note the order in which these principles have been abstracted: first, those having immediate impact on profit by making it worthwhile for a customer to give us his money in exchange for our product; second, the ones that identify the kind of information and the type of communication most likely to improve product value; and third, those that guide the smooth interaction among the people who contribute value most directly.

What Kind of Effort Do These Principles Govern?

The question calls for an operational definition of value analysis. The functional definition given earlier tells us what value analysis accomplishes and what it is supposed to do. The operational definition should tell us how it is done, what kind of effort is applied, and in what direction. Chapter 4 is meant to describe just that, but a chapter is not a definition. To capture the essential a definition must leave out the unessential and must therefore be brief.

The Operational Definition. Value analysis accomplishes its purpose by grasping opportunities and solving problems created by specialization, compartmentation, and rapid rate of change. So much for the definition; now let's examine it.

The opportunities and problems to which value analysis addresses itself are in the fields of information, communication, interaction, creation, and productivity.

Out of necessity, perhaps, modern business and industry have segregated the specialists, separated them by physical distance, by organizational barriers, and by mutually incomprehensible sets of acronyms. The resulting problems are dismissed as "failures in communication." More often than not the right information was never generated.

The principles of value analysis aim at better information, more accurate communication, and smoother interaction among specialists. Together, all 14 principles converge on what Koestler calls *The Act of Creation.*

Commerce and Industry CREATE Usable Goods. Creating aluminum ingots from the dust of the earth and magnesium from the salts of the sea are acts of creation no less useful than was purchasing grain in Egypt to ease the famine in Canaan at the time of Joseph and his brethren. Buying goods where they have little use and selling them where they have greater use is the creative aspect of commerce.

The Stultifying Lid on Imagination. In both commerce and industry the imagination of the work force has been inhibited by innumerable barriers and organizational no-no's. One way to increase productivity is to utilize the creative potential of the entire work force. Value analysis, therefore, conducts formal creative sessions to draw on the innovative potential of all key men and women associated with a given product.

I cannot pigeonhole ingenuity, resourcefulness, and inventiveness as isolated principles because they all contribute to the act of creation, and creation, the creation of usefulness, is the whole function of commerce and industry. Creation is more than a principle, it is a premise without which none of the principles is valid. The relationship is something like this:

PREMISES. Change, conflict, survival, creation, growth.

ECONOMIC PRINCIPLES. Consumer sovereignty, usefulness, also called the principle of function and performance, limited resources, balance and proportion, and economy, a fundamental principle in physics and mathematics as well as economics.

MANAGEMENT PRINCIPLES. It happens that these principles of management are also age-old principles of strategy. Here they are: direction—Clausewitz called this the principle of the objective—I used *direction* to avoid confusion between *objectives* and *objectivity*; adequate information, effective communication, interacting skills, the multidiscipline approach to reintegrating product value, selectivity, and concentration.

PRINCIPLES OF HUMAN BEHAVIOR. Direct motivation, objectivity, and responsibility or, better still, respect for responsibility.

Putting the Principles to Work

Application of the principles follows a sequence that conforms to the value-analysis job plan. We have to begin with consumer sovereignty to keep us from preaching unrelieved frugality. Once we reach agreement on the right of requisitioners and customers to their own judgment we can concentrate our efforts on the most promising portion of the project; then we have to do, with our own people, what it once took me 15 minutes to describe and the Japanese interpreter five words to translate.

He said, "Take the knives away from them."

The principles of direct motivation, interacting skills, and objectivity can be presented in a way that will convert hostility among disciplines and departments into what Professor Lamont calls *the fellowship relation*, ". . . the relation in which individuals stand to each other when they are pursuing a genuinely common good."

The other principles also follow the order in which they apply most directly to corresponding steps in the value-analysis job plan. Obviously, certain principles govern every phase of value work; others are narrower in application. The list that follows simply shows a convenient sequence of presentation.

Order in Which the Principles Are Applied

1. Consumer sovereignty	2. Concentration
3. Direct motivation	4. Interacting skills
5. Objectivity	6. Effective communication
7. Adequate information	8. Selectivity
9. Direction	10. Responsibility
11. Usefulness	12. Limited resources
13. Economy	14. Balance and proportion

Results

Applying the principles of value analysis makes the draftsman think as well as draw the picture, makes the buyer contribute ideas as well as buy materials, lets the girl on the assembly line dream of a better product as she assembles the old product, and provides an opportunity for all of them to exchange ideas with engineers, stylists, and financial managers. Tapping and mingling employee imagination fosters innovation and increases the true productivity of the entire labor force, not by finding better ways to make yesterday's products but by making better products for tomorrow.

Some Good Books

In 1969 Warren J. Ridge, manager of value engineering for the Analog-Digital Systems Division of the Control Data Corporation, came out with an extremely useful book called *Value Analysis for Better Management*, published by the American Management Association, New York.

If the 1961 *Techniques of Value Engineering and Analysis* by Lawrence D. Miles is the value-analysis bible, then the revised edition—McGraw-Hill, New York, 1971—is the new testament.

Another McGraw-Hill 1971 book is Art Mudge's *Value Engineering*. I have worked with Mudge and learned much from him. As a value specialist he is down-to-earth, pratical, realistic, and, above all, effective! He is the Director of Value Engineering Services for Joy Manufacturing Company.

Addison-Wesley, of Reading, Massachusetts, had the good fortune—also in 1971—to capture Edward D. Heller as their author on value. His book is *Value Management, Value Engineering, and Cost Reduction*. Ed Heller has contributed consistently to the value disciplines by his creative thinking and sound common sense. Like the others listed in this section, he is a successful worker in the field, being Manager of Cost Reduction and Value Control for the Convair Division of General Dynamics in San Diego.

II

THE NATURE OF VALUE

The average industrial employee is caught among the diverse values offered by the Prophets, the Devil, and the competition. He seldom has time to plumb the depths of economics and sociology for the morsel of wisdom that applies to the value he contributes to his company's products. He works hard, practices good human relations, develops a sense of political acumen, and makes moderate progress in his career. He wonders why certain coworkers who work less hard, are mean and ornery, make appalling political blunders, and produce fewer results, nevertheless get ahead.

Very often the answer is that the few results these men produce are the right kind of results. Moderate effort in the right direction has greater value than back-breaking effort in the wrong direction. That is the reason successful men seem to sail easily and pleasantly through their work. On the other hand, much useless and frustrating activity comes from working toward the wrong goals or from allocating effort among the right goals in the wrong proportions.

A sound understanding of the concept of value can help the man in industry to direct his efforts toward the right kind of results, to distinguish between more or less, on the one hand, and better or worse, on the other. More or less are matters of fact. Better or worse are matters of value. More or less require only effort, better or worse require judgment.

ONE DEFINITION OF THE CONCEPT OF VALUE

An idea that relates objectives to one another and to the cost of attaining them.

There is a lot more to it than that, for to understand value fully is to capture the essence of wise decisions made by successful men throughout the ages. We can hardly hope to do that, but we can approach it by striving to understand those aspects of value that are perceptible in our day-to-day work.

HOW VALUE WAS RENT ASUNDER

When industry went from individual craftsmanship to mass production, product value was partitioned and the product began to suffer under many masters. Making it work went to Engineering. Making it pretty went to Styling. Buying the ingredients went to Manufacturing. Finding the money and riding herd on it went to Finance. As craftsmen were forced to separate, their products lost value. Modern violins are inferior to the violins made by Antonio Stradivari; though abundant violins of reasonable quality have replaced scarce violins of superb quality—what, then, is the problem? First, loss of product integrity. The mix of desirable characteristics that the individual craftsman so carefully blended into his product is now contributed by a committee of departments; second, competition is sheer volume of production.

Competition

Japan is selling violins in the United States, Finland is selling "Spanish" guitars. Spain is selling motorcycles—competing with Japan and England for *our* motorcycle market. Yet competition is a source of strength and progress in the free world. The thing to do is not to bewail it but to beat it.

One way to beat competition is to recapture the high performance and fine quality of individual craftsmanship without losing the advantages of volume production. A value task group brings back together the people who were separated by the need to specialize. By recapturing the intimate interaction of skills which gave man his start in the first place, industrial teams—at the working level—make it possible for nations that cannot compete on labor and material cost to compete on brainwork and teamwork.

VALUE OF A FREE SOCIETY

The high labor costs in advanced nations result, in part, from a high standard of living. The people had the freedom to improve their lot. What was only a desire yesterday becomes an accomplished fact today and an actual need tomorrow. This very economic freedom offers a mechanism for successful competition. By helping convert dreams into reality a labor force improves its own technology. The man who was putting washboards together learns to assemble washing machines. When they become automatic, he must learn to wire the circuit. His son may already be wiring computers or he may be an engineer designing them.

NEEDS OR DESIRES?

Early value analysis was much concerned with eliminating the excessive, and rightfully so. The excessive is a baddie—useless by definition. It is bad because it excludes useful or enjoyable goodies. If the excessive is usually bad, the desirable is usually good. Ignoring this basic difference in matters of value—the difference between good and bad—can lead the unwary value analyst to lose the desirable in the belief that it is excessive. He then strives to provide "only the necessary function," to exclude the customer's desires, to work on "needs" alone. The question, of course, is who makes the distinction, and the answer lies in the nature of each society. The Spartans discouraged manufacturing and trade because these activities produced more than was needed.

From time to time human desires have been thwarted by persons and organizations who knew what was good for the other fellow. Examples are Lycurgus, Torquemada, Oliver Cromwell, and Cotton Mather.

Karl Marx advocated, "To each according to his need," but Lenin and Mao Tse-Tung both had to backtrack and give the people at least some of the things they desired. It is a political and economic reality in the Free World that the customer buys what he wants.

THE PRINCIPLE OF CONSUMER SOVEREIGNTY

The *McGraw-Hill Dictionary of Economics,** defines *consumer sovereignty* as follows:

* From the *McGraw-Hill Dictionary of Modern Economics* by Greenwald and Associates. Copyright 1961 by McGraw-Hill. Used with permission of McGraw-Hill Book Company.

The dominant role of the consumer in determining the types and quantities of goods produced by an economic system. The principle of consumer sovereignty is a key factor in the organization of production in a free economy. Each consumer purchase is actually the casting of "dollar votes" by which the consumer registers his desires in terms of goods and services.

Regimented consumers are the exception; they occasionally influence, but seldom determine, their own needs. Soldiers do not cast dollar votes on the ammunition they want any more than babies cast dollar votes on their breakfast food. A higher authority determines what is good for them.

Some governments have treated their people as though they were all soldiers or children. The purely military economy of Sparta, the rigid austerity of Cromwell's Britain and Cotton Mather's New England, and the state-controlled economy of East Germany all provide examples of meeting the people's needs and ignoring their desires. It has not worked too well.

Why?

Meeting needs but ignoring desires bars progress. Today's needs were yesterday's desires. To compete tomorrow—whether in business or defense—we have to work on today's desires.

THE RISE OF DISCRETIONARY INCOME

In the United States and in the countries of its major customers personal income has risen well above the lodging, clothing, meat-and-potatoes level. In 1789 the poor in Paris were storming the bakeries for bread. Today the poor storm the appliance stores. The forest of TV antennas over tenement houses tells us that the poor are not so poor as we think they ought to be. Even the most humble can pay for a little beauty in their lives.

In consumer economics this means that whereas yesterday's choice was a low-cost appliance or no appliance at all today's choice is between the old appliance and a *better* appliance.

Fortunately a scientific allocation of costs often produces a better product. Why not plan it that way? Improving the product at every opportunity is what it takes simply to stay abreast of competition. To move ahead calls for new ways to improve the product.

WORKING ON THE SIDE OF THE ANGELS

Is the value specialist an authority on all the functions of all the products of his company? Of course not. He simply gets the right specialists together at the right time, but he does not put them in the position of having to defend the functions of their products; otherwise the value effort would be feared as an attack to be endured, the recommendations as charges to be disproved, and the savings as a target for counterattack at the time of implementation.

As long as we regard product improvement and cost reduction as conflicting objectives we will have a hassle in the making. On the other hand, if the task group works as hard on improving the goodies as it does on reducing the baddies, it will place itself on the side of the angels.

My own experience, and that of the value analysts and value engineers whose work I have the pleasure of coordinating, is that if we stop saying, "Our job is to save money, not make the product better," the design engineers soon stop saying, "Our job is to make it work, not save money," and soon we are working together because product improvement and cost reduction are best done in relation to each other.

IS LOW COST *THE* MAJOR FACTOR IN SALES?

It *was* when Henry Ford stormed the market with his Model-T. For many people at that time the choice was either a low-cost car or no car at all. Most of the consumer's money in those days went for lodging, clothing, meat, and potatoes. There was very little left.

Even then, however, Professor Schumpeter (1934 translated from 1911 edition), writing on the importance of innovation in industry, pointed out that price cutting was mild competition compared with the competition of new products and new ways of performing old functions.

New models in consumer products may or may not be priced for less, but to hold their share of the market they have to be better. Let's face it: civilization is not going back from fine cars to Henry Ford's Model-T. Ford Motors is where it is today thanks to continuous improvement. So is Volkswagen. Volkswagen has added a gasoline gage, a lock for the left-hand door, synchromesh on all four gears, a more powerful engine, optional automatic transmission, air conditioning, and a whole new line of even more improved cars—a good example of how a small increase in cost can yield a great increase in utility. In a rising economy the customer demands better products at a price he can afford. A distinguished British scientist has put it much better than I can.

Cost Reduction and Cost Avoidance by Value Engineering

Journal of the Royal Aeronautical Society

February 1967, Vol. 71, page 94

Although originally conceived for the purposes of cost reduction, the objective of Value Engineering is now taken to be the optimisation of the product in respect of all its qualities, comprising cost, timescale, and the factors that make up performance, i.e., weight, drag, fuel consumption, reliability, maintainability, and so on.

Ewans, J. R., Manager
Economics and Project Analysis
British Aircraft Corp., Ltd.

J. R. Ewans, being an economist, obviously knows the difference between least cost and best value. Buying for the least cost may not return a good value, even when the necessary performance, quality, and reliability are assured. Here is an example from commercial electronics.

The mechanical support element of a solid-state component was selected for study. Cost was high in comparison with similar items which were manufactured in great quantities, but the cost was trivial in comparison with the electronic element it supported. It cost 43 times less. Say the completed item cost $44 and the mechanical element cost only $1.

"Why did you give us this to work on," the team captain asked. "If we throw the thing out altogether, the most we can save is a dollar per unit, $10,000 in 12 months. That would be less than any task group has ever saved in this division."

"Less than that," agreed the product manager, "because we won't be able to make 10,000 units in 12 months. We can't make them as fast as we sell them, thanks to this miserable little thing."

"Oh? . . ."

"Utterly undependable delivery. No vendor wants to make them the way we want them, in the quantities we need them, at the time when we need them."

The task group visited two vendor plants. There was no need to go further. "The way we want them" was not sacred. The task group recommended three different ways, one of which was acceptable to both vendors.

The minor change in configuration called for tooling which did two things: raised the unit price 10% and improved delivery 100%. The team's proposal was implemented immediately, with a resulting increase in production and sales.

In another division cost of slurry transfer pumps was $40 and they wore out at the rate of 10 per month. There was much down time and loss of product during replacement. The task group thought they could save 25% of the cost of the pumps, $100 a month.

"Peanuts!" was the team captain's indignant reaction. "I'm going to go back for a real project."

He returned immediately. "It is not the cost of the pumps that bothers them. It is the loss of product during replacement."

The team recommended an $80 pump which had four times the capacity. It cost twice as much but was four times as good. Actually, it turned out to be better than that. The linear speed of flow through the pumps was a major wear factor. Moving greater volume of slurry at lower speed virtually eliminated down time for pump replacement during production runs.

A leading bearing manufacturer advertises a bearing that costs twice as much but lasts twice as long as the old model. No advantage? Yes, indeed. Bearing changes are reduced from three to two.

BOTH COST REDUCTION *AND* PRODUCT IMPROVEMENT

in VA

A number of projects may be selected with the primary aim of yielding ready cash by reducing material and direct-labor cost; another group of projects may be selected with the primary aim of making a product more competitive in areas other than price; but *most* projects should be selected to yield *both* product improvement *and* cost reduction in relation to one another.

How this unmathematical levitation of more-for-less is accomplished will be explained later when we get into the value-analysis job plan, but here is an example:

You are told to reduce the cost of a piece of complex electronic equipment being made for the armed forces.

"Take out 20% of all direct costs," the boss says. "The price of everything has gone up since we bid on the job. We are going to lose our shirt!"

"Yes, sir! I'll put a value task group on every major assembly. The equipment will cost less and it'll be better."

The boss becomes a windmill of negative gestures. "No, no," he roars. "I don't want it cheaper and better. I want it cheaper and cheaper."

You object to the word cheaper, pointing out that value analysis doesn't ever cheapen anything, but he interrupts with a slowly articulated question: "Perhaps you prefer the word fired?"

"No, sir. I'll make it cheaper."

He looks at you with genuine pleading in his eyes. "Please, Carlos," he begs, "I know I told you to make the product better, as well as cheaper, but don't make it better this time, just cheaper."

Now, you cannot take 20% out of the cost of electronic equipment by savings on lock-washers, so you put circuit designers, electronic buyers, test engineers, and assembly-line foremen in the task groups.

Electronic circuit design is a value-analysis gold mine. While the designer is designing the circuit another electronic engineer, somewhere, is inventing better components.

You can take a brand new circuit, have a good electrical buyer call up a few vendors, and improve the circuit with better components—unless the circuit designer has availed himself of a value task group. If he has, he has designed a circuit around tomorrow's components and manufacturing methods, and that's what it takes to get the jump on competition.

After three days the task groups find more efficient components and better networks to get the most out of them. With all this new stuff reliability gets into the act. They look at the test data, scrutinize the qualification documents, and walk away shaking their heads.

Before you have had time to start worrying the phone rings. It is Himself. "Carlos," he growls, "I told you not to make it better."

"That's right, sir, and we haven't."

"The hell you haven't! You went and improved reliability. I have the figures before me right here."

This is a problem. You *did* try to follow instructions. "We couldn't help it, sir," you explain lamely. "With the new and simpler circuit, there are fewer parts to go wrong."

"What about the cost?" The voice is gentler now.

You grasp at a straw. "It was the cost that did it, sir. When we threw out the extra cost, the extra trouble went with it. You know, fewer parts, fewer things to go wrong; less heat generated, fewer component failures."

"Well," warns the boss, "see that you don't improve it on purpose. I want the most economical equipment that will meet specs."

But more trouble is brewing. The new components use up less power. Despite a smaller and lighter power supply, performance is better.

The phone rings. "Did you or did you not improve performance on—"

"Well, sir," you own up, "there was all this power that used to be wasted. It was cheaper to let it improve performance than to neutralize it. It would cost more to make it worse, but I might as well admit that the lighter power supply meant a lighter frame and—"

"I told you not to make it better. Lighter is better. I want it *cheaper*."

So you have to explain that when you took out the excess material, to make the equipment cheaper, the whole thing got lighter. It is now transportable by helicopter, and you have been up to your neck in Marketing people.

"Why Marketing people?" demands the boss.

"Well, since the equipment is lighter," you explain, "two other branches of the Service are buying it, which brings me to another problem we have created for you. Finance has run out of forms to invoice the new billings."

In Summary

You *can* make it both cheaper and better, and it might get you a convenient parking spot too.

WORKING DEFINITIONS LEADING TO A DEFINITION OF VALUE

CONSUMER. One that consumes; specifically, one that utilizes economic goods. A man home from work relaxing over a martini and a soldier using up ammunition are both consumers. The first is free, the second is regimented.

CUSTOMER. One that purchases a commodity or service, usually systematically or frequently. Customers buy for themselves, for free consumers, and for regimented consumers. The government buys PX goods for free consumers and uniforms for regimented consumers. A housewife buys olives for her husband's martinis and Pablum for her baby's breakfast.

WANTS (after Von Wieser). All human needs and desires, whether great or small, justifiable or unjustifiable, necessary or unnecessary. Bodily well-being, idle delights, and artistic pleasure may all be classed together as wants.

UTILITY. The capacity to satisfy wants. This capacity varies with performance, place, time, and quantity. Daniel Bernoulli pointed out the diminishing utility of each additional unit of currency to the man that already had a lot of money.

MARGINAL UTILITY. Utility viewed in the light of diminishing satisfaction derived from each additional quantity of a good.

Let us say that on a hot, dry day in July you have had the good fortune to be the umpire at a Little League baseball game sponsored by your local Baptist church. Exhausted, you stagger home in the hot sun and collapse at the foot of the icebox. Your loving wife kneels beside you and hands you a schooner of beer topped by a creamy head. That first draft of the cooling liquid is worth much more to you than the last draft of the fourth beer, and the last draft of the fourth beer is worth more than the sixth beer. It seldom takes more than six beers to recover from umpiring a Little League game, so the seventh beer is not worth anything at all to you at that time.

What is the sixth beer worth? The general theory of marginal utility, with its marginal rates of exchange and its indifference curves, is beyond the scope of this book, but I will say that marginal utility, when it answers the questions, "How much is too much?" and "When is too little worthless," is useful in relating customer satisfaction to the quantity or magnitude of certain benefits offered by a product.

Marginal Paradox. ". . . the paradox," writes William Smart in his *Introduction to the Theory of Value,*

(is) that the addition of items of goods is an addition of value only up to a certain point: if carried beyond, the Total Value falls; and if superfluity is reached, it disappears.

The total disappearance of Value, however, is almost never seen, because, at the worst, articles, however useless subjectively, have always the use of exchange.

The Little League umpire can trade the extra beer to an exhausted Scoutmaster in return for a patrol of scouts to serve as military police at the next Little League game.

Smart calls the paradox posed by marginal utility, "The Paradox of Value," but this term had long ago been pre-empted by the commentators on Adam Smith's *Wealth of Nations*.

Paradox of Value (Adam Smith's). The statement that extremely useful goods, such as water, have little or no exchange value, whereas certain other goods, such as diamonds, have great exchange value, though no use value.

William Smart comments:

Those who speak of diamonds having no use-value and of food as having infinite use-value, must be drawing their ideas, not from the life of men, but from the life of cattle.

One of the milestones in classical economics was the unraveling by Thomas DeQuincey and John Stuart Mill of Adam Smith's paradox of value. They agreed with Smith that free goods, such as water, may have great use value and little or no exchange value, but they denied that goods having great exchange value, such as diamonds, had no use value.

Mill challenged the puritanical interpretation of the word *use* in economics, maintaining that economics has nothing to do with the comparative estimation of different uses in the judgment of a philosopher or a moralist.

The use of a thing, in political economy, means its capacity to satisfy a desire, or serve a purpose. Diamonds have this capacity in a high degree, and unless they had it, would not bear any price.

Cardinal and Ordinal Utility. These terms really refer to two theories of utility measurement. The cardinalists argue that utility can be measured in degrees of satisfaction, whereas the ordinalists claim that only preference or indifference are true measures of utility. The argument is sterile.

In his excellent little book *The Measurement of Utility* Tapas Majumdar points out that the cardinal hypothesis is valid in some cases and the ordinal in others. He concludes his chapter on measurability with, ". . . from a priori considerations alone, the acceptance of one or the other of the hypotheses must remain, to the economist, largely a question of taste and temperament."

Utility in the General Sense. The quality or state of being useful or serviceable, as a farm tractor; fitness for a given purpose, as an engagement ring; aptness for a given application, as a parachute; beneficial for a given condition, as aspirin; timeliness in meeting a need, as a baby shower; suitability or convenience of location, as an oasis; in summary—completing the cycle—the capacity to satisfy wants.

The key words

useful	beneficial
serviceable	timely
fit	suitable
apt	convenient

illustrate the great breath and scope of the general concept of utility.

Elements of Utility

The specific properties that render a product useful or esteemed. These properties are usually descriptive of the effectiveness with which the product performs its function. Examples are performance, quality (including appearance), reliability, service, and opportune delivery. The utility of each of these elements, or their combined utility, can be quantified in dollars. How? A stranded motorist quantifies utility when he says, "A gallon of gas, right now, is worth five dollars to me!"

What does he mean by *worth*?

WORTH. An appraisal of the properties rendering a product useful or esteemed in the eyes of a *person;* a measure of such usefulness or esteem; the monetary equivalent of utility. From the West Germanic *wert* through the Anglo-Saxon *weorth,* this very English word reflects the direct approach to economic goods of Anglo-Saxon warriors, hunters, and farmers. By extension it has come to mean an appraisal of the effectiveness with which a product performs its function or a system accomplishes its mission. Either appraisal can yield a monetary figure that represents the customer's regard for the capability of a product to satisfy his wants. Closely related terms are merit and system effectiveness. The monetary connotation of the term worth makes it possible to quantify utility in the same units as cost.

Elements of Worth

When the elements of utility can be quantified in monetary units, we call them elements of worth so that we can relate them to their corresponding elements of cost.

FUNCTION. The dynamic aspect of utility. If capacity to satisfy wants is the utility of a product, actually satisfying those wants is the function of a product. Next to value *function* is the most important word in value analysis.

EFFECTIVENESS. The capacity to produce an intended result, such as meeting mission requirements or performing a product's function; often used as the physical counterpart of utility.

EFFICIENCY. The relationship of effectiveness to its cost in resources; the ratio of output over input.

COST. The unqualified term cost is used here to encompass total cost to the customer, including the supplier's cost and profit, sale price, and everything else the customer has to pay in order to acquire, use, enjoy, maintain, and dispose of the product, plus the time, effort, and risk of buying.

Elements of cost to the supplier can be reasonably identified by the usual cost analysis and cost accounting techniques. Examples of such cost elements are material, labor, tools, test equipment, plant rearrangement, travel, training, and the various expense and overhead charges.

Elements of Cost to the Customer

Price plus the costs of ownership, operation, maintenance, and disposal constitute elements of cost to the customer. All such elements of cost affect the competitive position of a product.

BEYOND SIMPLE DEFINITIONS

In order to define worth and function it was necessary to define utility and effectiveness and to distinguish between effectiveness and efficiency. All this should bring us close to a definition of value, but alas! it is not that simple. We are dealing with a relationship between sets of complex conditions.

Conditions Leading to Economic Value

Utility and Scarcity. Utility, or the capacity to satisfy wants, is a necessary condition for economic value, but it is not a sufficient condition. An antelope munching grass in an African mountain valley is certainly satisfying wants, as is the traveler quenching his thirst at a nearby spring. The grass and the water have utility of the highest order. So has the air that both antelope and traveler must breathe. Such utility, great as it is, however, does not engender economic value.

Why?

Because the tender grass, the spring water, and the pure mountain air *are free.*

Free goods have no value in the market place—people will not pay for them. Exchange value—the final measure of economic value—comes into being when a product, for which there is a customer, is made available to that customer by effort of some sort. This effort is required to overcome what John Stuart Mill called *difficulty in attainment:*

> That a thing may have any value in exchange, two conditions are necessary. It must be of some use; that is, it must conduce to some purpose, satisfy some desire. . . . But, secondly, . . . there must also be some difficulty in its attainment.

Scarcity, which presents such a difficulty, is generally considered a necessary condition for economic value. The broader expression "difficulty in attainment," however, is more truly descriptive. The Amazon region is teeming with monkeys and parrots. They are not scarce at all, but they *are* hard to get and therefore bring good prices in the market at Manaos.

Utility and Cost. We have seen that to have economic value something must be *both* desirable and hard to get. Wives are usually desirable, but they are not hard to get. So we do not, as a rule, buy and sell them. Porcupines, on the other hand, are very hard to get. But who wants porcupines?

When something desirable is hard to get for others and easy to get for

us, we are in business. We catch it, mine or manufacture it, and sell it for
the difference in the *difficulty in its attainment,* plus a profit for our
trouble. Now we are talking money. Finding uranium or baking bread
costs money. Such cost, plus the supplier's profit, is passed on to the
customer as price and becomes part of *his* cost.

Before he is willing to meet the cost the customer usually has some
idea of what the product is worth to him. Unhappy over being on foot
in a battlefield, King Richard the III offered his kingdom for a horse. It
is more usual, however, to quantify utility in the same units as cost.

UTILITY	is appraised as	WORTH
SCARCITY OR DIFFICULTY IN ATTAINMENT	are expressed in monetary units as	COST

COMPONENTS OF PRODUCT VALUE

The economic value of industrial products offered for sale depends on
the following:

1. Customers with money and unsatisfied wants—a market.
2. Utility to such customers—the product must suit the market.
3. Scarcity or difficulty in attainment—the product must be hard to
get.
4. Total cost to the customer—an inverse component of value. Given
difficulty in attainment, the customer wants to pay the least for overcoming that difficulty.

WHAT IS PRODUCT VALUE?

The interaction of these components provides us with a definition:

> *The value of industrial products is determined by that relationship of*
> *worth to cost which conforms to the customer's wants and resources in*
> *a given situation.*

The word *value* comes from the Norman French *valoir* and the Latin
valere, meaning *merit* or *worth.* It reflects the economic development
which followed Roman roads, Roman galleys, and Roman administration. As commerce and the circulation of money became commonplace
in the British Isles, *value* acquired the connotation of exchange in the

market place, while *worth* continued to mean usefulness or esteem independent of cost.

Worth is the simple concept. It becomes value when it is related to cost. Cost is, therefore, a necessary component of value in the same way that cost in resources, or input, is a necessary component of efficiency. The relationship between these two complex ideas, value and efficiency, is a most interesting area of exploration in value theory; both ideas relate performance of a function to resources expended.

Dollars and Words

The terms discussed up to this point are in common usage both by ordinary people—the real practicing economists—and by professional sociologists, economists, and mathematicians. Early value analysts would have liked to adopt such a simple and straightforward terminology. Had they stopped to do so, they might have had no value analysis to write about, but inspiration did not wait for formulation. The simple method of value analysis was originally described in casual language which served its purpose well. A discipline, however, is a body of knowledge that can be taught, and teaching calls for a careful look at the words we use. Whenever we ascribe obscure meanings to everyday words, we get into trouble. Calling the lowest *cost* of performing a function "the *worth* of the function" amounts to calling cost *worth*.

To the question, "Why not call cost *cost*?" the lame answer is, "Because some of us already call *value* cost. Value is often described as the *least cost* to achieve the necessary function."

The usual defense is, "Scientists make words mean what they want them to mean."

In the science of economics plain language is the best language. In his *Introduction to the Theory of Value* William Smart says:

> We are not at liberty to lay down new categories or even to give new names to economic phenomena. We have to take our categories and our vocabulary alike from the industrial and commercial world. . . . The theory of value, therefore, must begin with a careful analysis of what the word means in the mouths of ordinary people.

Professor Everett W. Hall followed this precept in *Our Knowledge of Fact and Value*:

> We shall start from common sense with our feet always placed on that firm ground (using) the categories of everyday thought, primarily as they are discovered in everyday language.

As often happens to new ideas, it is not the leaders but the disciples who dogmatize the words. The reason given is that any change in terminology would amount to throwing out all the work done since 1948; yet Columbus did not hesitate to question the thousand years of scholarship behind the flat-world theory.

What about the quandary:

We cannot call cost *cost,* because we already call *value* cost, so we have to call cost *worth!*

The industrial and commercial world offers an audacious solution: "Call cost *cost,* worth *worth,* and value *value.*"

MATTERS OF FACT AND MATTERS OF VALUE

What a product *is,* what it *does,* and what it *costs* are matters of fact. What it *should* be, what it *should* do, and what it *should* cost are matters of value. High speed and low speed, large size and small size, high cost and low cost are all matters of fact. The *right* speed, the *right* size, and the *right* cost are matters of value—when *right* is understood as *right for a given purpose.*

Greater and less are matters of fact, achieved mainly through simple effort in a given direction. The goals are a maximum or a minimum—the greatest or the least, respectively.

Better and worse are matters of value, requiring judgment to determine the directions in which effort should be applied to achieve the best value—the best *relationship* of worth to cost, not the most or the least of either.

ASPECTS OF ECONOMIC VALUE

The human urge to classify in order to understand has led men to define various "kinds" of economic value, but classification, as noted earlier, seldom changes the nature of what is being classified. Objects and concepts, however, have different aspects from different viewpoints. The discernible aspects of economic value do not differ from one another in a manner as clear-cut as, say, the front and side views of a house. The difference is often momentary, as in the plumage of certain tropical birds, where green seems to change into blue and then reveal flashes of red as the bird preens its feathers.

The first formal classification of economic values, which we owe to Aristotle, was colored by the Classical Greek's love of individuality and by their contempt for trade; hence the use of a sandal to be put on the foot was *more proper* than the use of a sandal to be given in exchange.

If the sandal cannot be put on the foot, it has no exchange value. If it can be put on the foot but has little esteem value because it looks ugly and smells bad, it also has little exchange value. If it can be put on the foot, looks good, and smells like rich leather but the town is inundated with such sandals, it has exchange value only in another market.

Exchange value, therefore, is affected by use value, esteem value, and market value. All interact, as Alfred Marshall put it, "like a number of balls resting against one another in a basin."

Before we discuss use value, esteem value, market value, and exchange value let us dispose of a red herring . . .

Cost! . . . Value?

This prickly concept, though rare on earth, is probably the dominant value in Hell, where the function is to provide the cost of sin. Cost value appears on earth as the result of ostentation or ignorance. To display their wealth some people buy costly goods and services. Thorstein Veblen, in *The Theory of the Leisure Class*, describes this valuation of cost as "pecuniary emulation," "conspicuous consumption," and "conspicuous leisure."

Lack of information sometimes creates cost value. To compensate for their ignorance of the product shy men buy the most expensive lingerie for their wives, taking a chance that it is also the best.

"Cost value" is a sad offshoot of the labor theory of value, which also proved to be in error. Adding the wrong kind of labor to a product, such as happens in the song, *Who Threw the Overalls in Mrs. Murphy's Chowder?*, adds to the cost but not the value.

Alas! many a casual reader of the value-analysis literature today equates "cost value" with the value of low cost. The two concepts are diametric opposites. Based, as it is, on the false assumption that the most expensive products are always the best, cost value is neither a good nor useful example of any of the values we try to enhance in value analysis. When we break a pencil in front of an audience to show that adding to the cost need not add to the value, we are not only demolishing the pencil, we are also demolishing the cost theory of value.

Use Value

The pivotal concept which gives rise to the other aspects of economic value is use value. As in the case of fuel oil, cinder blocks, and burlap bags, use value can appear alone, but it can also appear in combination with esteem value. Apples and peaches and grapes are food, a basic example of use value, but they are also beautiful and our wives use them for centerpieces. Tableware, clothing, furniture, homes, and even plumbing fixtures combine use and esteem value. The craftsman prizes his tools and the home manager, her kitchen utensils; so, the manufacturer makes them attractive and pleasant to work with.

Sometimes use value depends on esteem value, as in the wardrobe of professional entertainers or the paintings bought by an interior decorator to decorate a building. At other times esteem value develops from the excellence of use value. To many people a Rolls-Royce is beautiful—an object of esteem. Originally just another automobile, why did this car become a worldwide symbol of excellence? Consistently superior quality, dependability, comfort, and long life—all characteristics of use value.

The most costly production car on the continent of Europe is the Grand Mercedes.

Why do people buy it?

Prestige? . . . Perhaps, but also exquisite comfort, superb safety, and unmatched handling. One must admit that a Grand Mercedes offers great usefulness as well as distinction.

As I write this, I get a creepy feeling. Chills run up and down my spine and I hear a whispered conversation behind me.

"He is writing about things that give pleasure!" The accent and the stench of blood suggest Oliver Cromwell. "If he doesn't write about pure use value, let's behead him," continues the Lord Protector.

"No," answers a flat, nasal voice, "as a clergyman, I cannot shed blood. Let's press him to death with stones, as we do with Quakers in our colony."

That must be Cotton Mather. I quickly write down some examples of pure unadulterated use value: castor oil, mustard plasters, dental drills, long underwear, iron, bread. . . . And the wraiths vanish. The trouble is, however, that for every item of pure use value, such as a sawhorse, and for every item of pure esteem value, such as an emerald, there are many more examples of use and esteem combined, for "man doth not live by bread alone" nor can he eat emeralds.

If use value and esteem value appear so often together, why bother to make the distinction?

Opportunities in Combining Use and Esteem Value. In the same way that the proportions of saltpeter, sulfur, and charcoal determine the efficiency of blasting powder the proportions of use and esteem value often determine customer satisfaction. The customer may be an in-house requisitioner, designing an $800 hi-fi music system. The use of the system

is to please the ear. Why not please the eye as well? Why not, indeed!
To help sell the product, of course.

Esteem Value

The right feel and appearance of military equipment contributes to the
morale of the men and is a desirable form of esteem value. In the great
stream of American industry, which is, after all, the backbone of defense
supply, esteem value serves two purposes: it provides competitive advantage because a beautiful product is often less costly to make and
easier to sell than an ugly product, and in many instances it guides the
customer in the selection of functionally better products because in
good design as in nature form follows function.

Such principles of aesthetics as wholeness or unity of design, harmony,
simplicity, and economy also improve performance and reliability. The
lucid formulation which the scientist calls *mathematical elegance* stems
from the same source as the efficient design of the inspired engineer.
Both use their sense of aesthetics to let nature help them do what must
be done in the cleanest, simplest, most economical manner.

False Esteem Value. When an adornment does more harm than good,
it detracts from the value of a product. Examples are certain 1968 and
1969 automobile bumpers. One type was styled like a spear. It was
"sharp looking," all right—sharp enough to puncture the gas tank of
the car ahead, on collision. Another bumper was styled like a "W" or a
trident, combining the hazard of the central point with forward projec-
tions on each side which could pull a luckless pedestrian into danger
instead of pushing him out of the way. The function of an automobile
bumper is safety. When form fails to follow function, the product is in
trouble. Such a thoughtless grasp for disembodied esteem value often
elicits the comment, "They ruined it by getting fancy."

Subjectivity of Esteem Value. All economic value is, in part, subjective; but in the same way that in a democracy some people are said to be
"more equal" than others esteem value is more subjective than use value.
Use value is predictable. People put shoes on their feet, they all have the
same number of feet, and the number of people can be counted, but
there is no easy way of predicting how many of them will like brown
shoes.

Market Value

Understanding market value helps to avoid the trap of reducing cost at
the expense of customer acceptance. The customers speak through the

market, as James M. Roche, General Motor's chairman of the board, explained at the 1968 stockholder's meeting, "In the dynamic and changing market for new cars, our customers, through their purchases, tell us what they want."

Use value and esteem value are related to the physical properties that make a product satisfy the customer. Market and exchange value are related to its economic characteristics. The task of value analysis, with respect to commercial products, used to be to improve such economic characteristics as cost and delivery without detracting from the physical properties that make the customer want the product. The idea was to provide the same function for less cost.

Neither the same old functions nor the same old products can hold a market these days, much less break into a new one. Levy and Sampson wrote in their *American Economic Development*:

Twentieth Century America has seen a growth in the per capita income of its citizens perhaps unmatched during a similar period of time in the history of any people. . . . And, naturally, this tremendous increase in incomes has been reflected in the types of things bought by American consumers.

In his superb textbook on economics Paul A. Samuelson has this to say on our changing markets:

As people's desires and needs change, as engineering methods change, as supplies of natural resources and other productive factors change, the market place registers changes in the prices and the quantities sold of commodities and productive services.

Supply and Demand. Demand is, of course, engendered by customer wants and resources.

Some goods, such as the air we breathe, can be satisfied without expenditure of resources. People do not pay for them, and such goods have no economic value, but people do give their money in exchange for overcoming the difficulty in attaining what they want, can afford, and could not otherwise have. The economic process which makes such goods available constitutes supply.

Supply is provided at a price and demand is satisfied at a price. Price goes up as demand increases or supply decreases; it goes down as demand decreases or supply increases.

There is much more to it than that, of course, and it is with this "much more" that value analysis is particularly concerned.

We know pretty well why supply increases or decreases; the quantity of available goods and services is usually governed by matters of fact, but changes in demand are primarily governed by matters of value.

Valuation of Product Attributes. It is not size, weight, and quantity that determine the market value of a product but the valuation placed by customers on these and other attributes. Will large or small sunglasses be preferred by teenagers this summer? Will heavy touring motorcycles lose ground to lightweight scramblers and enduros? Will 6, 8, or 12 be the chosen number of flatware place settings three years from now?

Cultural Patterns and Consumer Choice. From Vilfredo Pareto to Thorstein Veblen and Kenneth Boulding, modern economists agree that social and cultural factors exert a major influence on a product's market value. One consumer chooses to use his credit, whereas another prefers to pay cash. Here we have a difference in what Boulding calls *preferred liquidity ratio.* When the whole market has a high preferred liquidity ratio, top-of-the-line products are harder to sell than when people cheerfully get into debt. Tradition or the desire to break with tradition, ostentation of wealth or ostentation of poverty, Christmas buying and spring cleaning are all examples of the many social and cultural trends that must be taken into account in analyzing market value.

Exchange Value

"The word, Value, when used without adjunct, always means, in political economy, value in exchange," wrote John Stuart Mill in the *Principles.* Exchange value is the final measure of product value. The customer answers all the questions we have posed in this section when he plunks down his money in exchange for a product.

The customer's money becomes return on sales. It is the major input into the general fund which provides wages, cash for reserves, dividends, interest payments, and payments for purchased materials and operating expenses.

To Summarize

The four most significant aspects of product value, from the standpoint of its analysis, are

USE VALUE
ESTEEM VALUE
MARKET VALUE
EXCHANGE VALUE

We should not assume that by thus classifying them we are separating them from one another in real life. Analysis takes things apart to understand them. To function they have to go back together.

OBJECTIVE AND SUBJECTIVE VALUE

The former chancellor of the University of Buenos Aires, Dr. Risieri Frondizi, can provide us with guidance in solving the objective/subjective problem. In *What is Value?* Frondizi writes:

> It cannot be solved if one persists in adhering to either one or the other position, or if one's major concern consists in affirming logical coherence to the detriment of the reality which one wishes to penetrate.

Then he asks, "Do values have to be necessarily objective or subjective?"

Professor Frondizi's little book—it is compact as well as concise—analyzes both the objective and subjective theories of values, screens out the sterile argumentation, and distils the valid contribution made by each theory. The conclusion is that the extremism of the two theories misses the point altogether. Economic value is, of necessity, both subjective and objective. A diamond, wandering around as a micrometeorite in outer space, can have all the objective properties that make it valuable on earth, but if there is no one there to value it it has no value. In the diamond fields of Brazil, on the other hand, there is someone who values a diamond—the enterprising prospector who makes his living from an occasional find. He values diamonds as his only source of food, tools, shelter, and possible wealth, but if the diamonds are not there or he does not find them they have no value. To borrow an expression from Alfred Marshall, the subject and the object are as necessary to the act of valuation as the two blades of a pair of shears are to the act of cutting.

VALUE AND THE CUSTOMER

Why the customer and not the manufacturer? In 1947 value to the manufacturer had already been thoroughly examined. It had been quantified as return on investment and profit and qualified by appraisals of customer acceptance, share of the market, and growth. Then the professional customers—the purchasing specialists—within a large corporation decided to analyze value from their own point of view as customers.

Value analysis, later followed by value engineering, was created to analyze purchased products and services and purchased design effort. It was, therefore, buyer-oriented. This buyer orientation, this concern with what the function does for the customer and for the ultimate user, has been the foundation of value analysis.

Customer-orientation poses the questions, "What does it do?" and "What is this worth?" and focuses thinking on the dynamic task of customer satisfaction, on the function.

Value analysis aims at value to the customer for the elementary reason that nearly all aspects of value to the manufacturer, from profit and return on investment to share of the market and growth, are dependent on value to the customer. But value to the customer—from the practical standpoint of the man in industry or the military buyer—has not been studied so closely nor measured so carefully as value to the manufacturer; yet value to the customer is the source of all industrial value.

WHO DETERMINES EXCHANGE VALUE?

The dominant position of the customer was brought home to me in a roundabout way. The head of the engineering support services at RCA's Missile and Surface Radar Division in Moorestown, New Jersey, had just hired me as administrator of value engineering. Leading me past a charming girl who turned out to be a very fine secretary and an even finer person, he then showed me into a pleasant office. "I hope you'll be comfortable here," he said. "Sit down."

I sat behind the desk.

"Open the lower left-hand drawer."

I did. There was a little sliding platform that covered it.

"Put your feet up."

Awkwardly, I put my feet on the little platform.

"Now relax."

This was too much. "How can I relax, on the first day on the job, *with my feet up*, when my new boss is explaining my duties."

"Those are your duties."

"Beg pardon?"

He pulled up a chair and sat facing me. "Go on," he insisted, "sit back, relax"—his voice took on a note of authority—"and *think!* You've been hired to *think*."

I could have told him he had the wrong boy, but you don't say that after you've quit the other job and the company has relocated you. Besides, the girl out there looked as though *she* could think. When God makes them right, they are not only pretty, pleasant, and structurally more than adequate, but they also can think.

My boss was saying, "I want you to think about value. Nobody knows very much about the value of industrial products and services. Hundreds of professors and consultants have their heads up in the clouds of macroeconomics, studying the Economy of the nation, with a capital E; more of them are working on microeconomics—the economics of the firm, of the company that makes shoes, for example, but nobody seems to understand just what makes up the value of the shoes themselves!"

He saw the pained expression on my face—I was trying to think.

"You don't have to come up with any ideas right away," he reassured me, "but I want to know how to put more value into the equipment we make. Bring in some Big Brains—operations analysts, systems analysts, some down-to-earth economists, our own psychologists, even philosophers if Security will let them in."

With all that help I was soon immersed in a 10-day seminar on the value of industrial products. We ordered dozens of books on value, argued about value, visited university libraries, and did research on value; then, one evening, I had a chance to get home. First I called the house to make sure they remembered me.

My daughter answered the phone. "Mamma went to the supermarket," she said. "Maybe you can meet her there."

Driving to the supermarket, I could hear the word *value* bouncing around in my head, begging for someone to define it. In the kind of mental haze that follows wrestling with a problem for too long I could see the actual word going from Aristotle to Saint Augustine to Kant to Adam Smith without finding a home. By the time the word reached John Stuart Mill I had reached the supermarket.

I found Maureen in that indeterminate area between food and housewares. "Look, Carlos," she said, holding up a red container, "this is a terrific value!"

Trying not to look superior, I asked her, "Just what do you mean by value?"

She looked up in surprise, not realizing the intellectual gulf that existed between us at that moment. "It's worth to me more than I have to pay for it—only thirty-nine cents."

Thirty-nine cents, I thought. How plebeian can you get! I tried to help her out. "Perhaps, my dear, you are thinking of the greatest good for the greatest number?"

"Carlos, this is rat poison!"

Trying to rephrase the sentence, I mumbled, "The greatest number of . . ." But it was no use. My masculine logic had been shattered. Why did we have to meet in that particular aisle anyway!

With genuine curiosity, I asked her, "What makes you think *you* can define economic value, when all the economists and philosophers—"

"It's very simple," she interrupted, opening her purse and pointing to the money inside, "I am the one who has the money"—cocking her head toward the manager's office—"that they want, so *I* define value."

And that's *it!*

The customer, who offers money in exchange for a product, determines the exchange value of that product by the amount of money he is willing to pay for it.

I related this incident in a panel discussion on *What is Value* at the 1966 National Conference of the Society of American Value Engineers. A panel member who had been defending what he called the *official* definition, protested, "Carlos is trying to impose the views of his charming wife on all of us."

The semantic confusion of that statement was pointed out by a member of the audience. My fellow panel member had confused the example with the concept, the map with the territory, the relation of an event with the imposition of an *official* definition.

It does not have to be my wife. The example was one of a customer in the act of buying. It may have been ill chosen, but I think not. The word *economics* comes from *oikonomike*—household management. It means, running a *polis*, a city state, the way your wife runs the house.

When a housewife is buying food for the household, she is practicing the art and science that economists write about. When she buys rouge and lipstick for her own use, it is another story—that is military equipment which often leads to the capture and pacification of *homo domesticus.*

To get back to economic goods, we can say that the customer is buying satisfaction of wants, which in turn depends on the utility of the product.

I do not say *our* product because *we* may be the customer. Value analysis was originally developed as a method for better buying in industry.

If the product costs too much for what the requisitioner wants it to do, value analysis improves the relationship between cost and function. From that point on it is only a step to using the method for improving not only the products we buy but also the products we sell; nevertheless, we cannot afford to forget that value analysis is customer-oriented. It was developed by a professional customer—a purchasing agent—for the use of other professional customers, his buyers.

The customer, in this discussion, can be an industrial buyer, a government agency, or a consumer in the marketplace. In each case he wants a fair return, in terms of utility, for the money he spends.

The special field of value analysis is the *relationship* between utility and cost. In order to perceive and improve this relationship, utility can be expressed in dollars, as shown in Figure 2-1. If something is worth

Calculating that the function is	he calls the ratio	because	and he calls the difference	which yields
worth – – – – $20 and the product costs – $10	a good value	it is worth twice (2.0) what it costs	a $10 gain	100% return
worth – – – – $10 and the product costs – $10	a fair value	it is worth what (1.0) it costs	no gain, no loss	no return
worth – – – – $ 5 and the product costs – $10	a poor value	it is worth half (0.5) what it costs	a $5 loss	50% loss
RATIO OF WORTH TO COST	QUALITATIVE RATING	INDEX OF VALUE	MEASURE OF GAIN OR LOSS	RATE OF RETURN

Figure 2-1 How the customer appraises product value.

$10 to the customer, he would like to buy it for $5. If it is only worth $3 to him, he will not pay $5. To make the sale we must either increase what the product is worth to him, reduce his cost, or both.

The index of value represents a ratio analogous to the ratio of efficiency in mechanics; both are ratios of output over input but the analogy stops there. In mechanics the output never exceeds the input; in profitable trade the return must always exceed the outlay.

The measure of gain or loss is the difference between worth and cost expressed in a common dimension. It is analogous to profit in business. It tells us *in numbers* how much the customer came out ahead or how much he lost, but it tells us nothing about the amount of resources invested. To find out whether the gain was worth committing those resources we must determine **the rate of return,** which is the gain or loss expressed as a percentage of the resources invested.

A study of the relationships illustrated in the chart reveals that there is much more to improving value than simply minimizing cost. It bears repeating that the task is to optimize value.

YET VALUE IS VALUE IS VALUE

The Good, the True, the Beautiful, all represent human values. Of these the Beautiful is the most nearly universal. Men have argued and shed much blood over good and evil and over what is true or not true, but even during an artillery duel, as the dawn heralds a beautiful day, gunfire slackens and enemies pause for a moment to see the sunrise.

The Beautiful is an intuitive, nonverbal indication of both truthfulness and goodness.

Far-fetched?

Let me use a down-to-earth example. Color TV is prettier than black-and-white. Why? Because it is truer to life. It conveys more information —brings us closer to the real world. Unlike black-and-white, it tells us that the peaches are ripe and that the girl picking them has blue eyes.

We could say, therefore, that the Beautiful reveals the True and the True reveals the Good, but intuition is more of a flash than a roundabout process. The Beautiful reveals the Good directly.

Beautiful weather is good weather—good for our health and spirit. Ugly weather is bad weather—bad for the lungs, bad for our aches and pains, and downright depressing. It brings to mind a turbulent sea hurling flotsam on a lonely shore, whereas beautiful weather evokes a picture of happy people playing on the beach.

There is something good about a pretty girl running along the beach. The balance and proportion of her healthy body, the smoothly flowing lines, the perfect harmony among the parts, all fill us with a sense of approval.

If she is smiling, she is even more beautiful. The smile is beautiful because it implies warmth and pleasantness which are certainly good. The beauty of her body tells us that she can perform all the tasks required of a female for the survival of the species—her children, obviously, will not have to be bottle-fed. She will be able to play with them all day long, clean house, cook, and still be a good companion to the happy man who shares in the production of such children.

That kind of beauty and goodness can no more be subjected to rational analysis than shrimp can be caught with a tennis net, for "the heart has its reasons which reason does not understand," as Blaise Pascal has told us. These reasons of the heart, governed by emotion and judged by intuition, determine esteem value, which in part affects the exchange value and market value of consumer products. Therefore, if people want a little beauty in their lives and they want it in living color, that is what we have to give them.

III

WHO DOES VALUE ANALYSIS

EVE STARTED IT

And when the woman saw that the tree was good for food, and that it was pleasant to the eyes, and a tree to be desired to make one wise, she took of the fruit thereof and she did eat.

Functions

(a) Good for food, (b) pleasant to the eye, (c) make one wise. The complex structure of masculine values is dwarfed by Eve's common-sense approach —first, food, then beauty, and finally wisdom.

Cost (To Adam)

In the sweat of thy face shalt thou eat bread, till thou return unto the ground; for out of it wast thou taken . . . Therefore the Lord God sent him forth from the garden of Eden, to till the ground from whence he was taken.

To Till the Ground! No more happy days of hunting and fishing, except by special dispensation of the Little Woman. With a triumphant grin, she generously gives her digging stick to the Man, announcing happily, "This is the beginning of the Neolithic. We are going to live in houses, weave cloth, and make earthenware pots. I'll be too busy to dig roots and gather berries and forever walk behind you carrying all your stuff while you have a ball hunting and fishing. Honey, you are going to have to learn to till the ground like the Lord said!"

Such was the Fall of Man—the change over from the hunting, wandering, and food-gathering economy of the Upper Paleolithic to the sedentary life of the Neolithic. Economic goods came into being, and in many instances the Little Woman became their custodian and beneficiary.

When we ask, "Who does value analysis?" one answer is . . .

THE LITTLE WOMAN

She started out with the apple, did not bother to pay the serpent a consultant's fee, and went on to develop the art of household management, which we call *economics* when it is applied to such masculine playthings as city-states, nations, and industries. While we try to fit her common-sense practices into the framework of college textbooks, she is teaching her daughters to play house, thus generating more practicing economists to keep one step ahead of us.

The Little Woman, as a household manager, practices intuitive value analysis. This book may serve her as a source of amusement to see how many of her age-old secrets we men have managed to steal. She may choose, however, to do formal value analysis in buying a car, a house, or in comparing the merits of independent and combination freezer and icebox units.

EVERYMAN WHEN HE MAKES AN IMPORTANT BUY

When a man buys something that is very important to him and he has to select one out of several choices, he often does so in recurring pulsations of anguish. Noting that one choice is superior to all others in performance, quality, cost, and appearance, he discovers that it is made overseas and that there is no dealership in the area. Another choice is sold only a mile away by a dealer with an excellent reputation for service; as a matter of fact, he makes most of his money from service because *his* product is always breaking down. There are three other choices. One is just about perfect, but the price is so high that necessary accessories will have to be bought one at a time later. The *one* choice that meets most requirements is horrible looking.

After suffering through endless cycles of comparison, Mr. Everyman finally makes his major purchase, painfully aware of its shortcomings.

Why all this frustration?

Mr. Everyman did not know how to determine the relative importance *to him* of the benefits he expected from the product. He had no way of comparing what each benefit brought him with what it cost him, and he did not know how to decide which benefits to give up.

When industry has spent millions of dollars developing a successful evaluation method and it is there for the taking, it seems a pity to pass it up. Mr. Everyman often learns about value analysis in his business; then he uses it to improve his personal life.

I know design engineers, purchasing directors, industrial psychologists, and financial managers who use value analysis to buy land for a new home, to design the house, to buy a pleasure boat, to design a lawn cart, and to help their sons select courses in school.

THE PROFESSIONAL BUYER ON HIS WAY UP

When Elissa, whom the Greeks called Dido, decided to found a new city, her brother must have hit the ceiling—a ceiling doubtless made of cedars of Lebanon, for their brother was King of Tyre, the great Phoenician trading center.

With the help of the fine Phoenician purchasing department, Elissa selected, from among Tyre's suppliers, the island of Cyprus, where she could pick up 80 maidens and a chaplain—a priest of the Mother Goddess. Then she set out with her followers, the girls, and the chaplain for the land of another of Tyre's vendors in what is now Tunis.

Tyre was the greatest trading city of the ancient world; and the adventure of Elissa illustrates the intelligence-gathering aspects of trade and the importance of the purchasing function as a source of information.

Of course, Elissa, being a good Phoenician, cleverly negotiated for land on which to build her city.

"We don't want a colony on our shore," the Libyans told her.

But she explained that she wanted to buy only as much land as an oxhide could cover.

Then she led the negotiating party to the high ground on a little peninsula, as though selecting the site for a tomb.

"This is the land we want," she told the Libyans, pointing to the ground at her feet.

They sold it to her and both sides swore by their gods to keep the bargain. Then she ordered the oxhide cut spirally into one extraordinarily thin and extraordinarily long strip which became the boundary of the New City, *Qart Hadasht,* in Punic, and Carthage to us.

The second part of the enterprise illustrates a self-defeating aspect of human interaction which the Buyer-on-His-Way-Up must value-analyze out of his operation—misinformation, subterfuge, and the tendency to be brilliantly clever in winning skirmishes and losing wars.

Carthage won her first, and many more, trading skirmishes until she became the wealthiest nation in the Mediterranean. Then she won most of the battles in the Punic Wars against Rome—all but the final ones.

If we are to believe the Greeks and the Romans, the Carthagenians were hated throughout the ancient world. For a great trading empire such an image suggests poor vendor relations and poor customer service.

Carthage, truly great for a time, used her trading fleets as a unique source of information on the activities of her rivals.

Value analysis, in the hands of the industrial buyer, offers a fast, direct method of integrating the three trading activities—information, customer service, and vendor relations.

Information

An order placer, we are told, simply converts pieces of paper from requisitioners into other pieces of paper which he sends to suppliers. It is not that simple, of course, for "converting pieces of paper" is a matter of communication, of deciding what is worth communicating, and of conveying the information quickly and efficiently. Order placing is not enough, however. A professional buyer must master the information, marshal the knowledge, and plan the strategy leading to a successful trade. He must, of course, use communications in the search for information as an aid to planning, as the means of conducting transactions, and to ensure successful results; but sound buying calls for much more than letting B know that A wants something and is willing to pay for it.

The subject of information, the difference between information and knowledge, and the not-so-simple topic of communications, are all discussed in Chapter 5. We are concerned here with how value analysis helps a buyer deal more effectively with the information, knowledge, and communications that constitute his stock in trade.

Every purchasing transaction starts with an item of information, either a requisition, an offer, or a discovery. The requisition usually leads to a search for suppliers or a selection among known suppliers; the offer and the discovery usually lead to a search for an in-house customer—someone who will benefit from the supplier's offer or the buyer's discovery.

Like any other improvement activity, value analysis must apply other than usual methods to achieve better than usual results. Instead of letting the input—the item of information—trigger a search for something else, the buyer applies an essential principle of value analysis and investigates the item of information itself.

The principle of adequate information requires that all informative inputs, especially descriptions and specifications, be up-to-date. Is this what the requisitioner wants *today*? The item in question may have met yesterday's needs. Will it meet today's and tomorrow's? The answer, of course, is up to the requisitioner, but he should not have to be a Noah Webster to explain what he wants to someone who works in the same plant. Face-to-face or telephone communication is one means of keeping information up-to-date.

Not only must the information be up-to-date in time but it must be accurate, it must be useful, and it should have a good signal-to-noise ratio. All this is usually taken for granted by routine requisitioners and routine order placers until the fabricated parts arrive on the assembly line and a foreman screams, "The holes don't match!"

Such an error, resulting from inadequate information or careless communication, was quite acceptable 20 years ago. The cost of rework and downtime was moderate in relation to the broad profit margin and slower production flow of the 1950's.

Today the profit margin per unit is narrower and the product flows through the plant much faster. Yesterday's downtime meant a little less profit. Today downtime and lost production may mean the difference between profit and loss. The modern buyer cannot afford to handle requisitions on the basis of faith, hope, and charity. He has to take positive measures to make sure that the supplier knows exactly what the requisitioner wants and that the requisitioner understands his own needs and what the market offers to meet them.

Isn't that going beyond good buying?

To compete today we have to go beyond what we used to do yesterday. Value analysis is sometimes hard to take because it forces us to face the fact of change. "I've been buying this way for 20 years" is no excuse for getting left behind.

Customer Service

As a professional customer the buyer understands what a customer needs to know. In the same way that he expects his suppliers to use what they know about their products to help him do better buying he can offer his in-house customers the knowledge of what will be printed in tomorrow's catalogs, the benefit of the supplier's efforts to come up with a better product to suit just such customers.

A good buyer should have at his disposal the product knowledge being developed by hundreds of white-smock scientists and technicians in dozens of vendor laboratories. *He* is their goal, their target, their customer. In many instances they will direct their efforts to design or formulate the very product that he will need tomorrow.

Vendor Relations

From the time when a warrior first traded flint spearheads in exchange for hardwood spearshafts trade has increased to the point that about half of modern industry's sales dollar goes into buying from other industries. At first it was a matter of necessity—you had to have flint or salt or furs and you gave something in exchange for them. Then it became a matter of convenience, specialization, and improved efficiency. Each specialist offers what he can do best in exchange for what another specialist can do best. The cold and weary hunter offers game to the hungry woodcutter and they both eat roast venison by the fire.

"Trade," they say to themselves, "is a good thing." The hunter does not have to carry a heavy axe and the woodcutter does not have to learn the art of stalking deer; but each has lost a measure of independence and autonomy; each is now dependent on a vendor. The vendor can be a solid staff to lean on or a broken reed to pierce the hand when it fails.

Vendor selection, therefore, carries the highest risk among all the activities of a buyer; yet his other functions, such as negotiating, recording transactions, expediting, and in-plant communication leave him little time for this crucial task. Here is an area in which value analysis helps the buyer to evaluate vendors on the basis of his requisitioner's needs. In the same way that he evaluates the relative merits of various products he can use the COMBINEX Aid to Choice and Decision—presented in Chapter 8—to select the most suitable vendor.

The buyer's trump card, of course, is value analysis in the hands of the vendor. Much poor procurement is the outcome of poor communications within the vendor's plant. A value-analysis program ensures that the supplier's engineers receive Purchasing and Manufacturing inputs early enough to do the product some good. It expedites delivery because much of the work that was done sequentially can be done concurrently, and it reduces errors by providing channels through which the vendor can go aggressively after the information he needs.

What if the vendor has no value-analysis program?

The buyer himself, or his company's value analyst, can organize a small task group—within the vendor's plant—to analyze the proposed procurement. Their questions will usually revolve around, "What is the best way to do this?" If, on the other hand, the question is, "What is it that we are really supposed to do?" the group should include people from the buyer's plant, hopefully including the requisitioner.

Major suppliers should be encouraged to set up a value program in their plants. The buyer, as "the customer," has a better chance of selling such a program than anyone in the vendor's own house.

In addition to practicing value analysis to improve his own buying, the buyer participates in task groups which serve the engineering or manufacturing functions. This gives him an opportunity to introduce procurement inputs early into the decision process, and it provides him and his boss with up-to-date information on the needs and plans of their in-house customers.

THE DESIGN ENGINEER

A key player in the industrial team that contributes to product value is the design engineer. There are ugly automobiles that work and sell; there are expensive automobiles that work and sell; but there are no automobiles that do not work and still sell—be they ever so beautiful and inexpensive.

The design engineer makes the product work by overcoming, through his ingenuity, John Stuart Mill's "difficulty in attainment." Both the words *engineer* and *engine* come from the Latin *ingenium*, from which we also get the word *ingenuity*, and ingenuity is the engineer's major contribution to exchange value, but his task is never so simple as that.

Looking at exchange value as a net to capture customer dollars, we can see that each knot, or nodal point in the net, corresponds to one of the components of product value discussed in Chapter 2.

Given customers with money and unsatisfied wants at node 1, the design engineer must create utility at node 2, overcome the difficulty in attainment at node 3, and keep the cost down at node 4.

With some exceptions, such as selling your soul to the Devil, exchange value involves an intricate relationship among the arts, the physical sciences, and the social sciences, all of which must contribute jointly to providing a product acceptable to the customer. What, then, is the engineer's role?

Scientist, engineer, beggarman, thief . . . doctor, lawyer, Indian chief? The engineer is a scientist-plus. The plus is economic feasibility, in exchange for which the design engineer trades off his formal contribution to pure science. The pure scientist is primarily an explorer into the unknown, a searcher for additional knowledge. His creative effort is in the field of discovery.

The engineer, although also an explorer, is primarily a pioneer and an inventor, putting exploration and discovery at the service of man. His creative talent is mainly in the field of invention.

Beggarman, thief? . . . Only in the sense of skillful scrounging, such as practiced by good sergeants in the army.

Doctor, lawyer, Indian chief? . . . These callings represent the diversity of customers an engineer must design for, not to mention the Army, Navy, Air Force, Coast Guard, NASA, and the Marine Corps. Visualize this array of customers and invite into the picture some of the missing ones—housewives, truck drivers, preachers, teachers, farmers, jobbers.

Would it make sense for an engineer, trained primarily in the physical sciences, to embark personally on the sociological task of interviewing all these classes of customer? Yet, he must please them. He is expected to apply the physical sciences to the service of man, and man is a social animal, so the experienced engineer teams up with specialists in the social sciences. Now, let's see how he finds them.

The Industrial Team

The engineer in industry actually creates exhange value but who defines it? Who establishes the requirements to make a product worth buying? The customer, of course. So industry has developed activities that specialize on customer wants—Marketing, Styling, and Advertising. Industry has also developed its own professional customers, its own professional buying activity—Purchasing. All these activities are studied under the social sciences.

Is the Engineer Caught in the Middle? Yes, in the key position which bridges the gap between the physical and the social sciences. Whether he is caught like a quarterback when the line collapses or whether he walks into the pocket to throw a pass depends to a great degree on the performance of his blockers.

How does he communicate with them? Originally, when our football developed out of Rugby, communication took place in the heat of the play, as it does in basketball today. Later, when plays were planned beforehand, a set of elaborate signals was used to designate the play, its direction, and the players involved. It worked for a while but not well enough. So the huddle became the rule, rather than the exception, and it proved to be a most effective means of communication, reducing the "signals" to a time delay and triggering device. The change was forced by the increased fluidity and dynamics of the game and by the need for greater flexibility and faster communication.

Industrial competition today presents the same challenge. Time— meeting and beating the schedules—is as important to the design engineer as improved performance or reduced cost. He has the choice of waiting for information or going out to get it himself. He often does the latter and often loses in accuracy and completeness what he gains in time.

The engineer rereads the specs, he telephones, takes trips, scans library shelves, and searches for the right people to answer his questions. All this time his sketch pad, slide rule, and lab instruments are idle. Instead of doing engineering work, he has been desperately seeking information that yesterday's methods do not provide in time for today's needs.

If he is astute, if he has been around the plant long enough, and if he has friends, he calls a huddle. Let the formal information flow in later; this time around he will meet face-to-face the specialists who have the data and skills he needs.

A committee? Hardly. The chairman of a committee is not supposed to take the initiative. A quarterback is. Committees were developed as parliamentary instruments to represent diverse interests. A football huddle, on the other hand, relates diverse skills to a common interest.

A Value Task Group

Who should make up the huddle? It should certainly include specialists in product assurance, cost estimating, purchasing, and in manufacturing, but it should also include a couple of other engineers—say one electrical and one mechanical—to evaluate financial, purchasing, and manufacturing information in engineering terms just as the other members of the group evaluate engineering information in business terms.

Such a systems-oriented task group exchanges and digests in a few hours the information that normally takes weeks to circulate in written, drawn, or coded form. More important, the information itself is more useful because its interaction is detected at once instead of being discovered later when something does not work, costs too much, or is not available in time.

Members of the task group ask each other questions that forestall time-consuming vendor inquiries, design-review objections, manufacturing problems, and marketing complaints.

What does the design engineer himself contribute? First, he provides direction, briefing the group on the problem or opportunity to be tackled and on the information he needs; second, he sets bounds within which the group must work; third, he exercises judgment in using or not using the information generated by the group; and finally he contributes the invaluable industrial ingredient of cool courage—that combination of judgment and daring which leads to a good batting average in risk taking.

As described up to this point, the design engineer and the task group have been getting along very well. Their relationship is one of trust, confidence, and mutual support.

Is this a real-life situation?

Remember, we said of the design engineer, "If he is astute, if he has been around the plant long enough, if he has friends, he calls a huddle."

To get the information he needs, when he needs it, a good engineer should not have to be astute, politically adroit, tactful, and persuasive. If he is talented and acts like a prima donna, the thing to remember is that prima donnas make a lot of money for opera companies and the thing to do is to provide both prima donnas and design engineers with a supporting cast.

In an industrial plant, however, members of the supporting cast often arrive on the stage not as a chorus but as a committee, each singing the aria of his own department.

Harmony in the Chorus

It is up to the value analyst to redirect the efforts of the supporting cast so that they can generate, exchange, and evaluate information with a view to its effect on their joint endeavor rather than their individual skills or departments.

Where friendship catalyzed the emergency task group described earlier, wholesome self-interest catalyzes the planned task groups of the value program. Each man gains advance information on what engineering is up to. This information gives his department more lead-time to prepare for what is coming. Whether he is a buyer, manufacturing engineer, or cost estimator, his suggestions reach the design engineer at the time when he can use them best. Specific purchasing, manufacturing, and financial recommendations enter the design cycle as initial inputs instead of reactions.

The greatest benefit to all participants in the value task group is in elapsed time. Today's joint planning at the working level forestalls tomorrow's panics, releasing management time to handle real uncertainty.

What has been said above about the design engineer as quarterback would apply to the responsible buyer if the task were primarily a buying task or to a styling specialist if the task were to improve customer acceptance by making a more beautiful TV cabinet.

VALUE ANALYSIS IN THE FACTORY

Can manufacturing technology improve product value? Of course it can! It has been doing it since the days of Taylor and Fayol. By dramatically reducing the cost of *making* the product manufacturing technology has provided the basis for the prosperity of Western Europe, the British Commonwealth, North America, and Japan. So much so that the cost of manufacturing—despite higher wages—continues to be an ever-smaller portion of total product cost. Some could say, "The job has been done too well!" But there are other worlds to conquer!

Improving the Product Itself

Instead of changing the factory to suit the product, how about changing the product to take advantage of improved manufacturing technology?

"Changing the product? That is not in our charter!" is the usual protest. "We are supposed to make it like the drawing, applying our ingenuity and resourcefulness to manufacture a product, no matter how weirdly it has been designed, how horrible it looks, or whether or not the materials have arrived. We are supposed to scrounge, to improvise, to *make* what somebody else has cooked up, not to help cook it up."

True or false?

Do factory personnel ever change a product because it cannot be made like the drawing? Of course they do. The facts of life simply move in and step all over the charter. As the result of a factory problem-sheet, a methods man and a design engineer get together to make the product more producible and often better all around. Yet year after year, despite this recurring opportunity, the manufacturing disciplines have searched only for better ways of making a given product, such as a tape-recorder, a cola drink, or an incandescent bulb, or of rendering a given service, such as electroplating, heat treating, testing, or packaging. Research in the factory has been limited to finding better materials and methods—better ways of making the product or rendering the service.

The questions "Should this be a reel-to-reel recorder?" "Should this be a cola drink?," or "Does it need to be packaged?" are not expected from the factory. With the exception of the suggestion program, factory-originated product changes are oriented toward producibility and toward savings in cost of manufacturing.

The same is true of most other industrial departments—strong concern with departmental performance and only a detached interest in the company's overall performance. Value analysis gives the man on the factory floor an opportunity to make his own specific contribution to the usefulness and desirability of the company's products and services.

For 20 years manufacturing engineers, quality engineers, line foremen, and factory cost estimators have been key participants in the value-improvement task groups serving Purchasing and Design Engineering. Today, particularly in Western Europe and Japan, value improvement task groups in the factory itself are considering the questions:

"Is this the best product?"

"Can we make it easier to repair and maintain?"

"Can we make it smaller, lighter, cleaner, or quieter?"

"Aren't we, after all, a part of the business?"

Changing With the Times

When Adam Smith wrote *The Wealth of Nations,* English industry was made up of weavers, potters, wheelwrights, cartwrights, blacksmiths, and similar craftsmen, operating out of their homes. Modest iron works, shipyards, breweries, and brick kilns represented "heavy" industry. Shoes were made by shoemakers, men's clothes by tailors, and furniture by cabinet makers. Smith's ideas on the division of labor laid the groundwork for the modern factory.

Half a century later, fascinated by the concept of the factory, Professor Charles Babbage left his chair of mathematics at Cambridge to become one of the first manufacturing consultants. If Smith and Babbage brought the craftsman into the factory and divided the craft into discrete tasks, Frederick D. Taylor and the Gilbreths, Frank and Lillian, divided the task into discrete movements. Then Henry Gantt projected the operations against time in his planning charts. All of this is now accepted as part of modern manufacturing, yet at each step of the way hundreds of voices rose in protest, demanding, "What is the difference between this and just plain good management?"

The difference, of course, is the difference between good and better, between stagnation and change. It is a difference embodied in the concepts of adaptability and improvement, improvement in the worker's physical output and, thanks to Gantt, in management's attitude toward the worker (see A. W. Rathe).

Mary Parker Follet (see also Metcalf and Urwick) showed how the worker's judgment and ideas could become part of his productive output. Meanwhile, Henri Fayol, the eminently successful French industrialist, had done for the front office what Taylor, the Gilbreths, and Gantt had done for the factory floor. Finally, Lillian Gilbreth tied the whole thing together.

What else was there to be done?

Something for the function of the product as well as how to make it, and it was at this key-point in time, 1948, that Lawrence D. Miles joined the ranks of Taylor, Fayol, Gantt, Follett, and the Gilbreths to take us one more step forward along the road to better industrial management.

ANY DEPARTMENT MANAGER

For contrast let us first look at one of the few purchasing agents who do not use value analysis. It is one of those days when he feels like the traffic controller in a crowded airport. Unwanted, early shipments, preceded by their invoices, are circling overhead. Scheduled shipments, with all ground facilities waiting for them, are nowhere in sight. Departures of revenue-producing planes are delayed pending the arrival of the late shipments. Ugly weather is moving in and everybody is burning precious fuel.

How is such a situation usually handled?

Our harassed purchasing agent rushes around from expediting delayed shipments, preventing early shipments, and explaining factory shortages, back to expediting more late shipments which cause new factory shortages, etc. . . .

With a telephone receiver in each hand, he barks at one of his buyers, "Can't you see that I have no time now for vendor selection? Use the people we know, the people we can depend on."

And he goes back to undepending on his unselected or poorly informed vendors.

Is this primarily a purchasing problem? Hardly. It is a management problem—the fire-fighting syndrome. The department manager has been thrown off balance by an emergency and is now fighting a rearguard action; new emergencies have been created by the first emergency. Had he had time to fight the fire and still handle the job, he might now be in control of the situation. As it is, the Goddess of Chance is kicking him in the pants, retaliating for the overload.

What overload?

When we leave to chance something that should be our job, chance resents the imposition and clobbers us. A defensive posture leaves too much to chance. The only way out is to regain the initiative, but to regain the initiative one needs reserves. Of these reserves, the most important is time. Time cannot be stored but it *can* be reserved.

Creating Time Reserves

To avoid the fire-fighting syndrome a department manager reserves time in the future to handle specific problems. He schedules this time just as he would schedule business trips, but he also schedules time to provide for the unforeseen, thus creating a general reserve of time.

Value analysis plays a minor but useful role in comparing the cost of holding out reserves with the cost of being caught flat-footed, but it plays a major role in providing the necessary time by eliminating unnecessary effort. Immeasurable time is lost in industry because of the lack of current, accurate, and opportune information. The methods for getting, filtering, and using information described in Chapter 5 liberate enough hours in a department manager's day to provide this time-reserve.

Better than fire-fighting, however, is fire prevention. The department manager can prevent many a fire by allocating time and effort in proportion to the importance of the results expected.

Value of Timeliness

The question, "What should come first?" is not easy to answer. Does it mean "first" in the sense of before and after or "first" in the sense of more important? Could it be "first" in the sense of cause and effect? Chapter 1 discussed the relation of time to causation. Chapters 8 and 9 relate it to the decision process. In this chapter we look at how the successful department manager allocates his own time to ensure his business survival and growth on the one hand and the effectiveness of his department on the other. Good department managers have been analyzing the value of their own time long before value analysis was formalized. Doing it methodically simply makes the process easier.

What To Do and What To Put Aside

I remember working for a small company that designed and manufactured excellent telemetry receivers. Today I would be called a project engineer and I would be in charge of one or two projects. I was then called a production engineer and had forty-eight (48!) projects, except that we called them "jobs." The company had a strong incentive program. On Friday mornings we were told, "Those of you who get the work out on schedule this week will get to keep your jobs." This was reality. The company depended for its existence almost entirely on technical know-how and Spartan discipline.

I would arrive at work one hour early on Monday morning, look at my files, and ask myself, "Now, which of these jobs is going to get me fired Friday?"

And I would work on that one, thus "minimizing the chances of maximum loss"—the minimax method of game theory. This method serves when it is absolutely necessary to gain time, but it will not take a manager from the defensive into moving forward.

The department manager, using value analysis, first segregates the tasks of his shop according to the kinds of time that govern them. The checklist in Table 3-3 has been developed for this purpose. The manager uses it to allocate his own personal time, to monitor certain major tasks, and to train his subordinates in analyzing the value of *their* time and in pacing their progress.

To make sure that the user does not miss any key points that list is pretty thorough, but only really pertinent information should be entered. Many of the reminders are often not applicable to a given task. The ones that are must be handled in the manner most convenient to the user. Dates and elapsed time are simply written in. *Worth* and *cost* can be described in words and so can *value*—the estimate of worth related to cost.

A simple rating scale, with no zero point and ranging from 1 to 5, can serve to sort out rough appraisals. Table 3-1 shows how the scale is used.

Table 3-1   Words into Numbers

What It Is Worth to Me	Rating	What Is the Cost	Rating
Slight worth	1	Very low cost	1
Small worth	2	Low cost	2
Moderate worth	3	Moderate cost	3
Good worth	4	High cost	4
Great worth	5	Very high cost	5

The numbers can then be used to express the ratio of worth to cost as a common fraction. Converted to a decimal, this fraction yields an index of value which ranges from 0.2, *a very poor value,* indicated by the fraction 1/5, all the way up to 5, *a very good value,* indicated by the improper fraction 5/1. By eliminating repetitions we get 19 different value indices (see Table 3-2).

Why not call 3 *a very, very good value,* 4, *an outstanding value,* and 5, *an utterly terrific value,* instead of lumping them all together under *a very good value?* We are not lumping them together. We are simply re-

Table 3-2 Numbers Back into Words

Index of Value	Verbal Description
5, 4, 3	Very good value
2, 1.66, 1.5, 1.333, 1.25, 2.5	Good value
1 (unity)	Fair value
0.8, 0.75, 0.666, 0.6	Poor value
0.4, 0.333, 0.25, 0.2, 0.5	Very poor value

fusing to *un*lump the gruel of things as they are into an imaginary pap just because it is easier to measure.

Five levels of value are enough to rank most tasks in order of importance; for instance, a timely start which promises substantial advantage (4) at low cost (2) would have a value of $4/2 = 2$, *a good value* according to Table 3-2. If the advantage is only *fair* (3) and the cost is high (4), the value is reduced to $3/4 = 0.75$, *a poor value*.

Paragraphs A, B, and C of Table 3-3 identify those tasks that are keyed to the *flow* of time, tasks that must be done before or on a certain date or dropped altogether. Some of these tasks may be trivial and others very important, but each can be put aside just so long and no longer.

In deciding what to do first, you automatically reject what *not* to do first. The natural tendency is to put aside the unpleasant tasks and do the pleasant ones first; or, conversely, to get all the ugly ones out of the way and clear the decks for the ones you like to do. Planning your work on the basis of personal preference is perfectly sound, provided you have three or four months' wages in the bank, an updated résumé, and your family is adjusted to relocating. Otherwise, it is better to select and reject tasks on the basis of their urgency and importance. Here is an example:

You are an anthropologist and the chief of a tribe of Huitoto Indians invites you to visit the fierce Jivaros. He looks at his five wives to select companions for the trip. Their eyes brighten and they all look at him expectantly. He selects two and thus rejects three. The rejected girls pout.

You are appalled. He has picked out the two oldest women, leaving three luscious babes to pine away in loneliness. Discretely, you remonstrate.

"Blanco" the chief explains patiently, "the Jivaros live far away, too far to carry guns. We will have to live on fish. The three girls are too young. They haven't learned how to catch fish with their hands. They don't even know how to find the vines to squeeze into the stream and stun the fish."

By this time the thoroughly competent elder wives have picked up the gear and are on the march. In this case necessity rather than personal

Table 3-3 Checklist for Analyzing the Value of Time

Task _____

Person responsible _____ Location _____

A. Timelines as Opportunity
    Earliest feasible starting date _____
    Most *timely* starting date _____
    Deadline before "missing the boat" _____
Worth of timely start _____
Cost of timely start _____
Value of "striking while the iron is hot" _____

B. Timing for Best Sequence
    Must precede _____
    Concurrent with _____
    Must follow _____
Worth of right sequence _____
Cost of right sequence _____
Value of right sequence _____

C. Time as Duration
    Working days required _____
    Starting date _____
    Completion date _____
Worth of completing on time _____
Cost of completing on time _____
Value of completing on time _____

D. Setting Priorities
    Should be done before _____

    _____

    Concurrent with _____
    Should follow _____
Total worth _____
Total cost _____
Economic value _____

Strategic value _____

Risk in doing this job _____
Risk in not doing it _____

preference determines selection and rejection, but . . . wives are a time-sensitive resource. To know what to do first you have to know what you can put aside and for how long. That is the purpose of the first part of

Table 3-3, and Huitoto Indian chiefs have no such checklist. They go off on long trips, accompanied by their sole-source travel wives, and express great surprise and indignation when they find out on their return that the girls they left behind did not stay put.

The entry, *Strategic value,* in Table 3-3, may have intrigued the reader. It refers to the value of a particular task in contributing to the career goals of the person using the chart as related to the career goals of his superiors, his subordinates, and his co-workers. It is a measure, really, of human feasibility, such as the likelihood of the wives-in-training being there when the Huitoto chief returns.

The department manager not only improves the value of time by planning tasks in the most profitable order but he adds directly to the value of time by allocating it in large, usable portions. The most elementary value analysis reveals the loss from frittering away time in useless driblets.

Each driblet of time is as nothing in accomplishments compared with what the same interval could do as part of an uninterrupted stretch. All related minor tasks, for example, can be scheduled ahead to be done together, provided they are worth doing at all. This technique not only concentrates the use of time itself but it makes possible the concentration of attention on a specific kind of activity, keeping the channels of memory recall open, the scales of judgment balanced, and the decision gates trimmed to handle that class activity.

Successful concentration of attention eliminates extra set-up charges in time, energy, and peace of mind, which we have to pay every time our thoughts have wandered or been driven away from the task at hand.

The principle of concentration, as applied in value analysis, calls for assembling in a task group the skills best suited to cope with a given problem or to exploit an opportunity. The group then concentrates its attention on that aspect of the project likely to yield best results for the time, skills, and manpower available. In this sense the principle is the same as the principle of concentration in military science, which is discussed under *The Element of Strategy* in Chapter 9, but the general *principle of concentration* has broader applications. It enhances value by removing the unessential and undesirable as in mining, in which concentration means the removal of the less valuable part of the ore, or in chemistry, in which concentration means the removal of the foreign and unessential in order to increase the strength of a compound or solution. Winnowing, which separates the wheat from the chaff, screening, filtering, centrifugal separation, and distillation accomplish the selective accumulation of properties useful for a given purpose. By throwing out the inert or the harmful these techniques concentrate desirable elements in greater strength and purity. This form of concentration by elimination can be likened to cost reduction, in which value is improved simply by eliminating unnecessary cost; but there is a form of concentration that does not depend on elimination. The condenser lens in a slide projector simply concentrates light by redirecting it to where it will do the most good. This is an example of concentration in space. Light that would have gone elsewhere is concentrated in a smaller area. I am indebted to my cousin Sancho for another example. Sancho was traveling with a theatrical company, two of whose members had just been married. Early in the morning, following the wedding night, he was awakened by the girl's voice in the room next door. She was saying, "Now, I'll get on top," and after a pause, "No . . . better *you* get on top," and then, "I know, we'll both get on top."

There was a pounding on the wall and he heard his own name. "Don Sancho, Don Sancho," the girl was calling, "come and help us. We're both on top!"

Motivated by chivalry and scientific curiosity, Sancho rushed to the aid of the young couple. He found them sitting together on top of a suitcase they were trying to close. Adding his weight to theirs, he completed the necessary concentration of three persons on two square feet and the suitcase was finally closed. Had the whole theatrical company sat on the suitcase, one at a time, nothing would have been accomplished.

The yellow blinking lights placed around construction work on roads provide an example of concentration against time. So does an electronic flash in photography. The principle is the same. Power from the battery cannot provide the necessary intensity, not if the little electrons are marching to their work in the usual manner; but an obstacle is placed before them, the metal plate of a capacitor. It says "Wait here." Well, it doesn't actually *say* it, but it is placed in a definite wait-here-attitude. Just beyond, across a gap that determines the length of the pause between flashes, another metal plate is waiting in a welcoming posture. The electrons pile up on the first plate. When there is no room for more, the last one to arrive shouts, "Come on, let's go!" and they all jump together, providing a brilliant flash. Weak energy has been concentrated against time and released in powerful, usable surges.

In summary, the principle of concentration can be applied to assemble, bring to bear, or converge forces, activities, or resources, to strengthen by elimination of the nonessential, to condense the desirable, to concentrate in space, and to concentrate against time.

Purchasing's Objectives

Before planning to do something, we have to find out what is worth doing. A task group, representing a department's in-house customers, as well as the doers who provide services for them, can give the manager a new insight into his department's objectives.

The task group starts out by finding out the true mission of the department. The function of a purchasing department, for instance, may not be "to buy things" as Mark Shepherd, Jr., president of Texas Instruments said in the January 11, 1968, issue of *Purchasing Magazine*.

I think that the purchasing mission is to be part of a larger system creating profit and growth. . . . I want them to become more informed and influential.

He expects Purchasing to influence design through its knowledge of tomorrow's materials and to support Marketing by pointing out new business opportunities and reporting on trends in the market place.

In the same issue of *Purchasing* (1/11/68), Roger S. Ahlbrandt, president of Allegheny Ludlum, emphasizes Purchasing's contribution to the company's public image, and C. B. Burnett, president of Johns-Manville notes Purchasing's contribution to the company by value analysis and by researching new ideas and letting the company try them out.

The department manager is well aware of the stated and expected goals of his operation. A task group will help him delve into his department's unsuspected but useful contribution to company objectives.

To follow the Purchasing example a little further, if Purchasing supports Marketing and the Design House and is also a powerful influence in Public Relations, it will have three customers who, if treated right, can strengthen its position. They should be represented in the task group which serves the department manager in determining the department's objectives and the goals that measure progress toward those objectives.

Distinction Between Goals and Objectives

Marvin Bower, managing director of McKinsey & Company, in his book, *The Will to Manage,* recommends distinguishing between goals and objectives. The dictionary does not help very much but Bower does. Grasping at the subtle points of difference barely implied by dictionaries, he characterizes goals as attainable and time-related and objectives as more general and longer lasting, "typically timeless."

The goal, then, is what we can do and will certainly try to do this quarter, this year, or in the next five years. The objective is what we are striving for all the time.

An engineering department's objectives may be to innovate, design, and develop gas turbines, advance the state of the art, and stay ahead of competition. The goals would then be measures of specific progress toward those objectives in a given period of time—typically a year:

1. Initiate a program to discover a medium for gas-cooled reactors with gas turbine cycle. Lab equipment to be installed in June. Prototype of turbine designed by year-end.

2. Design and develop a closed-cycle turbine and a clean-exhaust turbine for automotive use. Drawings to be released in September.

3. Introduce our helium-gas turbine at the Nuclear Industry Fair.

4. Find out if Brown-Boveri, Ruston, or Sola are doing anything with composite materials in turbo blower blades.

Marketing, Purchasing, and Manufacturing people can provide the manager of this engineering department with first-hand information on what he needs to know in setting goals jointly with his people. Add a financial officer to the task group, and the "worth to the firm" (Odiorne) can be compared with the cost of attaining a goal, thus arriving at its value. Such a task group can ensure that each goal will be set on the basis of its contribution to higher objectives and not simply because progress is easy to measure.

Measures of Performance

Analyzing the value of many current "measures of performance" is one of the hazards of value analysis. Simply changing a few of them can turn an operation around from loss to profit.

Is that bad?

It is, if it embarrasses people who have been living with the losing measures for years and who are the ones who have to implement the change.

A classical example is the factory overhead rate. A manufacturing engineer gets a value task group together to analyze a complex assembly. It turns out that all the complexity is the result of hasty, step-by-step design required by a tight schedule.

The design engineer agrees that the assembly can now be cast in one piece, eliminating countless operations. He prepares new drawings, tests the prototype, and approves the change.

"Can't do it," says the plant manager. "Takes out too much direct labor. My overhead rate would shoot from 125 to 145%. The home office is holding me down to 130% factory overhead."

The factory overhead rate, of course, varies all over the picture—all the way up to infinity—in the ideal fully automated plant. Many people still think, however, that a low figure is good. It is neither good nor bad, but it is *not* a measure of efficiency. A value task group, headed by a senior financial officer, can develop measures to create rather than destroy incentive.

Most negative measures date from the days when a manager's function was to keep the people busy, frightened, and unhappy. With the exception of a nursery school, a tour ship, or a prison camp, "keeping the people busy" is not a modern business objective. Activity, for its own sake, does not yield a profit.

The value task group therefore searches for those measures that refer to activity alone. A low factory overhead rate simply means that many people are working and few people are supporting them. What the mix should be must be determined by productivity, not by the belief that more brawn and less brains are always better. As we shall see under the *principle of balance and proportion* in Chapter 7, it is not the quantity of each but the mix that counts.

Another measure of activity alone is the number of drawings that come out of a drafting department. The function to *convey information* becomes *produce paper*. To look good, according to this measure, the chief draftsman sighs in disgust and orders his people to disperse the information among the largest possible number of drawings. Much better measures are the accuracy and quality of the drawings and the time-cycle from sketch to delivery.

Some purchasing departments are measured on the number of orders placed, the number of salesmen interviewed, the number of phone calls made, the number of business trips taken, and the number of conferences attended. The sheet I have before me—happily from a company in another industry—also lists the number of salesmen refused interviews and the number of overtime hours worked. Not one word on the results achieved. Measuring people on how busy they are is like measuring a hunter on how many shots he fired instead of how many ducks he brought home. Today's Purchasing people are pretty good hunters; they are proud of their ducks. It is better to measure them on what they do for the company than on how busy they look.

Perhaps one of the most absurd measures imposed on Purchasing is cost per purchase order. To reduce this figure the purchasing manager can follow one of two courses: reduce the number of his people—often at the expense of the economy and quality of what he buys—or simply break up the large orders into many little ones! The cost to the company is higher, but what he is measured on, the cost per purchase order, is lower. He smarts under the strain to his integrity or he does more than smart and suffer, he gets a value task group together. "Fellows," he tells them, "we have pooled your specialized skills to develop measures of performance that do something for us," and they do!

Here are examples of such measures:

Level of Quality. A measure of the performance, quality, reliability, and other requirements of purchased materials in terms of acceptability as related to a base period. This is a measure of results, of the worth obtained by buyers under specified limitations of cost and lead time. It serves as an incentive to question unreasonable quality requirements and to meet or surpass logical ones. It does something for us.

Timeliness of Delivery. A measure of the opportune arrival of purchased materials achieved by comparing actual with expected delivery dates. It rates the effectiveness with which buyers plan and control delivery. The date expected can be based on standard lead time for given materials or classes of materials. This is no harder to arrive at than standard cost, but it must be adjusted to the vendor's current capabilities. The date expected is really an appraisal of the buyer's ability to forecast the arrival of the materials. To improve this index he must draw on his forecasting experience and develop a sense of anticipation. This measure develops greater foresight. It does something for us.

Expected Outflow of Cash. This is a measure of the ability of the Purchasing Department to anticipate the cost of purchased materials at the time of payment. It keeps buyers on the alert against price increases, and it highlights savings achieved by ingenuity and resourcefulness. A reasonably accurate prediction of the outflow of cash for purchased materials helps Financial Management to plan the allocation of funds. It does something for us.

Prevented Loss of Cash

Good buying may well spend more money to improve delivery, quality, or service when the improvement is worth more than the added cost, but by and large good buying saves the company money. The savings report *is* a measure of results. Cash that would have gone out the door is kept in-house, but the savings must never be at the expense of required quality, delivery, or the supplier's fair profit. Well, then, savings at the expense of what? At the expense of monopoly, unnecessary requirements, waste, and needless uncertainty. This measure, whether we call it *prevented loss of cash, cash conservation*, or the "savings program," usually identifies the major categories of savings so that greater effort can be applied to the most effective ones in each given situation. It helps buyers to reduce the outflow of cash profitably and to improve their savings techniques. It does something for us.

THE GENERAL MANAGER

Productivity of labor in North America is lagging behind skyrocketing wages and material costs. Making goods at today's costs and selling them at tomorrow's prices may appeal to the optimist who has spent his last dollar on a money belt to keep the earnings. But there is a catch to it. Predicting sales in an economy subject to "fine tuning" can lead to feast-and-famine inventories. All too often the feast turns to famine because today's hard goods are perishable! Nobody pays full price for left-over annual models.

The general manager can analyze the value of having annual models against their cost. The annual model serves to identify a product as "a little better than last year's," which is what many a customer wants in a rising economy because *his* wages have been rising as fast as prices. Other benefits are the simultaneous cut-in of improvements and the marketing momentum of the model-year pageantry. The cost is primarily that of instant depreciation of warehouse stock at the end of the model year. Value analysis provides a means for cost-to-benefits study of this situation, with specific techniques for avoiding self-deception, wishful thinking, and the misinformation that timid subordinates offer their superiors to make them feel good now and suffer later.

The incident I will relate now, however, illustrates a more general and truly fundamental application of value analysis, combating the profit squeeze by giving the customer less of what he doesn't want and more of what he does want. The incident takes place on the Inland Waterway in Florida, and the customer is a charming and fertile young woman.

A breakfast food manufacturer, whose name I will not disclose because he was literally crying in his beer on the stern of his yacht, said to me, "In the old days, it took only plain, good management to stay ahead. People had to eat. You produced good food, and you sold it to them."

"Aren't you doing that now?" I asked.

"No. It has to pop and crackle." He sighed as though we were in a mausoleum instead of balmy Florida.

"Pop and crackle!" I rolled out the syllables. "If that's what the customer wants, it may be cheaper for you to produce it—one of the ingredients is air." But he had started me off on one of my lectures. "Plain good management," I went on, "continues to worry about finding better ways to make a given product, but that battle has been won and—"

"Just a minute," he protested. "What do you mean *won?* Productivity per man-hour of labor is rising *faster* in our South American subsidiaries than it is in this country."

"From how much to how much?" I asked.

He pondered audibly, "Well, let's see. . . . Productivity down there was not too high to begin with."

"Didn't they have a long way to improve from?" I suggested. "Aren't they still fighting the production battle on the factory floor? This is a battle your people have won."

"Maybe you got something there." It was a grudging concession. "Productivity in our stateside plants *was* high to begin with. I guess it can't get much higher too fast."

It was Sunday. I was not working. I was drinking beer and sunbathing, but I couldn't resist the opportunity to say that value analysis is a method for directing the necessary change in emphasis, from "how to make it" to "what to make"; the change in battleground, from the factory floor to the market place; the change in the use of manpower, from muscle labor to brain work; and finally, the change from using the hands of some people and the brains of others to using the brains, ideas, and enthusiasm of everybody in the organization.

Sulking, he looked despondently down at his beer. "What do you want me to do?" he wailed, "tell my manufacturing people that productivity can't be improved because they did too good a job?"

As his wife arrived with two frosted steins of beer and a bowl of shrimp, he complained at large, "I know what Carlos is going to say! He is going to say that my job is not to make fine energy-producing cereal, but to package air into crinkly little lumps that crackle and pop in the bowl—."

"And make a satisfying crunch, crunch sound as my children chew on them," his wife added, "so that you can sell more and pay those rising costs."

" *Your* children," he exploded, "are *my* children—"

"Presumably," she acknowledged demurely.

"And if you are feeding a competitor's breakfast food to *my* children, you are feeding them—"

"Language!" she warned. "But I won't argue, not so long as you think you know what the customer really wants," and she started to leave.

I protested. "Sit down, charming customer, and tell us what the customer really wants."

She slipped gracefully into a deck chair.

"The last thing our little bundles of energy need is *more* energy," she said, half to herself, stretching out her suntanned legs. "I have enough trouble hauling them down from the rigging as it is. What we mothers need is a bulk-producing cereal with whatever sound effects are required to make the children eat a lot of it—without getting fat—and then feel drowsy enough to be quiet for a little while."

In the same way that my friend discovered a typical customer right in his own family many a general manager can find priceless customer information scattered among all the people who work for him. Each, in his own specialty, can contribute a little toward making the product more desirable. Value analysis provides a method for integrating all this Engineering, Purchasing, and Manufacturing brain-power with the Marketing and Styling functions in order to make sure that the entire organization contributes to making the right products for today, instead of making yesterday's products right.

John Stuart Mill, as noted in Chapter 2, postulates two conditions for exchange value. The first, remember, is that the product must "be of some use, that is, it must conduce to some purpose, satisfy some desire," and the second that "there must also be some difficulty in its attainment."

The manufacturing industry has profitably dedicated scores of years to reducing the "difficulty in attainment" and to reducing the cost of manufacture and the amount of "direct labor" that goes into a product. The time has come to work on the *first* condition of exchange value, the usefulness, purpose, and desirability of the product, what value analysis calls *the function*.

THE VALUE SPECIALIST

An architect, a shipbuilder, and a machine designer, if their products have customer acceptance and yield a profit, must certainly practice value analysis. After all, isn't value analysis "what we try to do all the time?" It is, and *some* of us succeed, intuitively or otherwise.

Intuitive Value Analysis

The buildings of Sir Christopher Wren and the ships of Joshua Humphreys are fine examples of intuitive value analysis. Perhaps the best example is the way Robert Stephenson value-analyzed his father's invention, the locomotive, into such a simple and efficient propulsive engine that it remained unsurpassed for a hundred years. Wren's buildings grace London to this day, cherished by one generation after another, and Humphreys' frigate, the *Constitution*, turned out to be such a fine ship that it has been preserved as a national relic.

What have such accomplishments in common? What general principles govern their attainment? How can those principles be applied to improving product value? In seeking answers to such questions, the value specialist—whatever his title—has his work cut out for him.

Task of the Value Specialist

Lawrence D. Miles was the first to identify those characteristics of intuitive value analysis that can be methodically applied in business. He started out with the methods of sensing, collecting, appraising, and using information. Not information on buildings, ships, or locomotives but information on the functions that such objects perform and information on what resources it would take to accomplish them. This and the other steps of the method are what this book is about.

Information. Once a project has been selected, getting the right information to the right specialists at the right time is the first service rendered by the value analyst. This rule applies, whether the analyst is working with a task group, whether he is part of an independent value analysis section, or whether he is working on his own. In the latter case he may have to collect and disseminate the information personally, though it is practical sometimes to identify the key specialists and put them in direct communication with one another.

Analysis. Identifying the product's function, breaking it down into the benefits it provides, comparing their relative importance, and relating the benefits to the cost of attaining them, is the second service performed. The experienced analyst can usually identify and define the functions of a product almost automatically. When working with a diversified task group, however, he must teach the participants how to do it. When the function is complex enough to be broken down into the benefits it provides, the relevant specialists must participate in the study, either in the task group or by individual contact with the analyst.

Innovation—Alone or in Groups? So much has been written about group creative sessions that many a value analyst feels at a disadvantage when working without a task group. There is no evidence, however, that a man alone is inherently less creative than he would be as part of a group. What happens in well-selected groups is that there is a great pool of relevant information that provides grist for the creative mills of the individual members. The groups also provide stimulating discussion and cross-pollination of ideas.

How does the analyst, when working as a loner, achieve cross-pollination? The same way that bees do. They flit from flower to flower, carrying pollen from one flower to the other and engaging in stimulating discussion with each flower. In touching base with the specialists, the lone value analyst participates personally in the interchange of ideas instead of being only a catalyst. It may take more time, but often it is the best way to do it.

Not Either Or, but Both And

The various ways a value specialist can work are not mutually exclusive. In a small plant he may personally work on two projects, providing service for a five-person task group and, at the urgent request of the controller, planning a two-week, profit-oriented workshop to bolster next quarter's earnings.

Implementation. In all cases implementation of value-analysis proposals is the decisive service, for the value specialist is not measured on seminars held, persons trained, or brochures prepared but on results.

WHO ELSE USES VALUE ANALYSIS?

The reader will have noticed throughout this section that in the industrial applications of value analysis the buyer, the design engineer, the manufacturing engineer, and the department manager have each used a value task group to bring the benefits of other skills to their own operations. Participants in each group represent, at one time or another, nearly every industrial specialty in the plant and in the field.

Engineering, Manufacturing, Purchasing and Cost Estimating are the most frequent participants, but Finance, Quality Assurance, Marketing, Styling, key customers, and key vendors also participate, depending on the task at hand.

We know what each participant contributes—his own special skill and the readiness to provide up-to-the-minute information on a custom-made basis for the users of that information, but what does he get out of it? He gets information in return, not only about the present project but about other opportunities not suspected until the group got together; *and* he establishes a network of contacts among key activities in the plant, contacts that will prove useful to him for many months to come, not to mention specific techniques for working successfully across the barriers that often separate industrial disciplines.

We might say, therefore, that all participants in the value task groups apply value analysis to improve their own day-to-day operations.

IV

HOW IT IS DONE

REQUIREMENTS

The practice of value analysis in business and industry requires (a) a knowledge of the method; (b) an understanding of the nature of value; (c) the ability to work with diverse specialists in gathering, appraising, and using information; (d) knowing how to analyze a product's function; (e) knowing how to improve the relationship between function and cost or, to put it another way, between product worth and product cost; (f) the ability to innovate on schedule; (g) skills in evaluation, comparison, and choice; and (h) dexterity in implementation.

Acquiring the Necessary Skills

Like swimming, fencing, and chess, formal value analysis is best learned by doing it. Since most people do informal value analysis all the time, simply formalizing one's day-to-day value appraisals provides a good start toward the theoretical foundation.

As an aid in this process, the material in this book has been arranged in the step-by-step sequence of the value-analysis job plan.

Face-to-Face Communication

The reader already knows that no single discipline in industry contributes *all* the value in a product. He also knows that the contribution of each individual discipline is usually pretty good and that the area most subject to improvement is that portion of value contributed by the *interaction* of disciplines. A simple and direct way of improving this interaction is by face-to-face communication.

Touching Base with the Specialists

The experienced design engineer goes to the factory floor to make sure his design is producible. He returns by way of Purchasing to find out if his materials can be procured. When he first received the assignment, he touched base with Marketing and may even have talked to the customer. The first thing he was told about the new product was, "It has to sell for $499.95"; therefore he has to discuss his plans with Cost Estimating. This trail is individually blazed by most design engineers as soon as they discover that the traditional methods of handling information simply cannot keep up with the rate of change in modern technology. So-called "information" or "data" is indeed transmitted, stored, and retrieved instantly, but is it *information?* This question is taken up in Chapter 5. The problem we are facing now is that of the design engineer who must develop a personal search-pattern to get the kind of information he needs when he needs it.

In the same way the experienced buyer touches base individually with those specialists who are most directly affected by his purchase. Because value analysis originated in Purchasing, today's professional buyer is usually familiar with the method. In touching base with his requisitioner, he does not say, "I want to make sure I understand what this part is," but "I want to make sure I understand what you want it to do."

The buyer also has friends in Cost Estimating and in Production Control. In talking to them he asks "Why?" much more frequently than yesterday's buyer who, after all, was expected to buy what he was told to buy. Today's buyer understands the relationship between value and information. He not only meets face-to-face with his requisitioner and his supplier but he may take the two out to lunch together. He also touches base with the quality and reliability specialists, and after he has touched base with everybody he collapses into his chair, kicks his shoes off, and stretches out in semiconsciousness. He is too tired to place the order that day.

THE VALUE TASK GROUP AND OTHER APPROACHES

Much individual value analysis starts with the excursions described above. It is certainly better than the traditional "coordination" in which the left hand did not know what the right hand was doing, but it is time consuming and requires a powerful lot of walking. A more economic and often faster way is the use of value task groups. These groups reduce leg work by putting all the legs under one table; and they save time by the concurrent exchange of data.

First Let's Look at the Other Approaches

The reader may be ill at ease when working with a group, and he may be able to sail easily through the excursions described in the preceding paragraph, in which case he may do better with the individual approach. But it isn't so simple as that. There are four major factors that govern the way to analyze product value: (a) the temperament, skill, and experience of the analyst, (b) the nature of the business, (c) the nature of the product, and (d) the timing of the effort.

The Analyst as a Technician. The do-it-yourself analyst must have considerable experience in actually doing value analysis, as distinguished from teaching it or directing it. He must also have substantial skill in one or more of the major disciplines that contribute to product value in his company. Generally he is or has become curious, inventive, and resourceful.

The Analyst as a Catalyst. To be primarily a catalyst the analyst must have infinite patience and understand, in depth, those aspects of value analysis that have proved successful in promoting effective interaction among diverse specialists. He is usually articulate and persuasive.

The Analyst as an Instructor. If he is to specialize as an instructor the analyst must be a master of those aspects of value analysis directly concerned with integrating the constituents of product value. Whether he realizes it or not, he is usually well grounded in economics and human motivation.

Do these types exist in the real world? Not in their pure form—none could function without some of the qualities of the others—but they do exist within each successful value analyst. The question is, which facet should be uppermost when facing a given task?

Snoopy, the little dog in Charles Schultz's comic strip *Peanuts,* can show us how effectively our personality patterns can be fitted to varying conditions. Snoopy has a confrontation with the detestable girl, Lucy, who meets him on a narrow trail. Crouching low, Snoopy advances. "This is a Bengal tiger," he tells himself, "a Bengal tiger stalking his prey!"

Lucy mutters, "There is something wrong with that dog," and gives him a wide berth. Snoopy showed that facet of his personality best suited to the occasion. It takes more than that, of course, for . . .

There Is No Substitute for Solid Know-How. Being articulate, tactful, and inventive is not enough to do value analysis; on the other hand, understanding the method and being able to apply it, can convert the quiet technician or the scholarly economist into an effective analyst, for success in value analysis depends more on technical competence than it does on exhortation.

The analyst's background and temperament—related to the business and to the product—should determine the choice between the individual and the group approach.

Advantages of the Individual Approach. When working without a task group, the analyst must interact personally with many diverse specialists, thus building a rich reservoir of product knowledge and an increasing understanding of the causes of poor value.

Analysis by the individual requires less preparation on the part of the analyst and is easier to start than analysis by a group. Substantiation of facts and record keeping come about almost automatically, and the specter of collective decision making does not even raise its head.

The Independent Value-Service Group. Quite different from the multidiscipline groups is the special group that provides value services. In the same way that airframe manufacturers have a structures group and a weights group, many companies have a value group. A major portion of the value analysis under my direction has been done by just such a group. This group, attached to the home office of a major commercial-products division, serves three large plants and a number of small ones. Its members use the individual approach in projects assigned to them. Working together, they also conduct excellent profit-oriented workshops with multidiscipline groups.

The Task Group Approach

My particular interest in this approach resulted from reading critiques prepared by the participants. Here are some examples:

"I found out who *They* are and why they do what they do. It turns out that I am part of this *They* everybody beefs about."

"We learned the problems of other departments and how we had been contributing to those problems. They learned about ours too."

"For once, I found out what information the other guy really wanted and I could give it to him on the spot."

"I am a technical man. This workshop helped me acquire some business sense."

One of the earliest and most valuable suggestions was, "Drop dead. Soon. In a deep hole. Why? You showed me up, you rat-fink! That's why."

I learned from this one, first, that one can write effectively in sentence fragments; and second, that the writer of such a recommendation does not expect full compliance and may be willing to negotiate. This benefactor, for such he turned out to be, signed his recommendation and provided a phone extension so that I could "report progress." The third, and most important lesson, was that it is stupid to hurt someone's reputation needlessly, a lesson that has stood me in good stead ever since.

I had to undo the damage by answering management's question, "Why wasn't this done in the first place?" Before answering such a question, a manager should try to answer it in his own mind, for the answer may be embarrassing to him and to his boss.

Causes of poor performance and needless cost are often "lack of information" and "failure in communication." Identifying these shortcomings is not enough. They have to be prevented, and for this purpose the value task group is an excellent tool.

Selection of Task Group Projects

This paragraph is really an intrusion at this point, but it is necessary because the reader should know that the projects must be selected before selecting the people who will analyze them. To use typical planning language, "We select the *what* first and then decide *who* are the best people to do it."

In Chapters 5 and 10 we discuss in greater detail how to select projects, but at this time we have to consider how the groups are tailor-made to suit different types of project. It bears repeating that the word *product* is used in this book for *products, services,* and *operations*—pretty broad coverage! But then, *value* is a broad concept.

Analyzing New Material Applications

In the mid-fifties, when the extraordinary properties of Teflon were barely recognized, Nems-Clarke, a manufacturer of top-quality electronic and nucleonic equipment, got a task group together to take advantage of the new material. The group included Joe Rupert, head of the model shop, George Dellinger, industrial engineer for the electronic assembly lines, George Triby, buyer, Harley Peter, electrical engineer, and Aldo Mautino, mechanical designer. Note the diversity of disciplines. Teflon-coated wiring was adopted for all areas in which high parts density created heat problems; Teflon-coated slides were installed in the racks of electronic cabinets, and the lubricity of Teflon was utilized to ensure the smooth and silent movement of the gamma-ray source-carriage inside the massive shield of the radiac calibrators.

A Group to Search for Future Products

An electronics plant was suffering from defense cut-backs at a time when the space program was going into high gear. Management organized a value-analysis task group made up of a market analyst, an applications engineer, a physicist, a circuit designer, and a cost accountant. Their chore: "Study our skills and capital equipment and find the space-age products we are best fitted to design and manufacture." Note the diversity of skills and their particular suitability for the task at hand.

A New-Product Task Group

Sometimes a technical advance calls for complete product redesign. Changing the design is often easier than replacing skills and changing machines to make the new design. It is easy for a theatrical producer to change a boy's part into a girl's part, but it is not practical to change the boy into a girl. It is often impractical to change existing machinery and skills to suit a radically different design.

In the theater the producer finds the boy another part. In industry we have to find something else to be made by the surplus skills and unused machines.

To answer the question, "What can we sell that is *made* like our old product?" a marketing task group was set up. After considerable research the group came up with a number of good recommendations, one of which showed exceptional promise, except for the drawback that the materials necessary to make it did not exist. When this drawback was pointed out by the materials manager, one of the marketing men admitted, "Oh, we knew that all along, but we are always sitting in on your value-analysis task groups. Now *we* could use a little help. How about a value task group to find—"

"To find nonexisting materials?"

"Exactly! They don't have to exist today. We can't go into production today; but it would be nice if you made them exist tomorrow."

"Who do you think you are talking to, God! . . . ?"

Everyone bowed deeply.

Their reverence produced a value-analysis task group that included two subcontract managers, one to find out which of the division's major suppliers were capable of developing the required materials in the quantities, time, and cost we needed and the other to work with sister divisions to arrange for their help in the search; a corporate staff attorney or his representative to guard against reciprocity and monopoly; the head of Equipment Development to get a preview of the materials he would have to deal with; and the procurement specialist on the particular class of materials.

A Design Support Group

Value analysis of routine new designs and of existing products is by far the most common application of the method. This applies in both make and buy situations. The group is meant to provide information for the designer from disciplines other than his own. The *designer,* as the word is used here, may be a design engineer, an architect, a fashion designer, or a pharmacologist in a drug firm. He needs fresh and useful information on (a) customer wants, (b) customer resources, (c) worth of the product in terms of use, (d) worth of the product in terms of esteem, (e) cost and procurability of materials, and (f) cost and producibility in manufacture.

Figure 4-1 can serve as an aid in the selection of the specialists that should be included in a task group. If the "difficulty in attainment" of the product, which induces the customer to part with his money, is a matter of making it work, then value in use is the designer's main contribution. Since he is the man whom the group serves, he should not form part of the group. The whole idea is to release his time and energy for design decisions by getting specialists in their own fields to generate information that he would otherwise have to scratch for. But he cannot leave the most important skill unrepresented. He needs someone of his own specialty and of his confidence to make sure that the information generated is what he actually needs.

If the product is mechanical and the designer is a mechanical engineer, he needs another mechanical engineer in the task group.

If the product is a drug, the pharmacologist in charge needs another pharmacologist in the task group.

TEAMS	SKILLS	Marketing, Styling, Quality Assurance	Design Engineering, Product Engineering	Industrial Engineering, Manufacturing Engineering	Cost Estimating, Cost Accounting	Purchasing, Production, Control, Subcontracting
Team No.	Projects					
1						
2						
3						
4						
5						
6						
7						
Workshop Location:						

Figure 4-1 Selecting a balanced combination of skills.

Some Typical Value Task Groups

The listing that follows shows the make up of task groups for particular products in a variety of industries.

STANDARD STEAM TRAP, PROCESS INDUSTRY. Process control foreman, piping engineer, steam engineer, capital equipment buyer, cost estimator.

TV CABINET, CONSUMER TV INDUSTRY. Acoustic engineer, furniture designer, styling specialist, assembly-line foreman.

TV CHASSIS, CONSUMER TV INDUSTRY. Assembly-line wireman, electronic engineer, mechanical engineer, time and motion specialist, electrical buyer, mechanical buyer, cost estimator.

ANTIBIOTIC, DRUG INDUSTRY. Physician, biochemist, pharmacologist, chemical-process engineer, chemical buyer, quality specialist.

Team or Committee

Committees were originally developed, within the parliamentary process, for the purpose of reconciling conflicting interests; hence the statement, "A camel is a horse designed by a committee." Nobody is completely happy with it but the majority accepts it.

Since a committee is meant to represent interests, it usually meets in an atmosphere of polite conflict. Committee meetings are not, therefore, the pleasantest chores. How often have we heard the expression, "I've just come from a miserable committee meeting. We talked for two hours and didn't get a thing done."

Contrast that statement with the cheerful, "We had a lot of fun!" so often heard after a value-analysis session.

Is There Such a Thing as a Group Decision? Certainly. The stampede of a herd of cattle. An audience leaving a burning theater. A lynching. Hardly good decisions. Good decisions are usually made by individuals, not groups. So-called "collective decisions" often represent an attempt to evade or share responsibility. A military commander, and not his staff, makes the decisions in a campaign. The military staff, nevertheless, is an invaluable adjunct to the commander's decision process, providing continuously updated information, contingency plans, and a choice of options.

As distinguished from a committee that meets to reconcile departmental interests, a value task group meets for the interaction of departmental skills independently of departmental interests.

After a value task group has determined the best combination of product benefits for a given level of resources it may well reconvene as a committee to make sure that the interests of the various departments are safeguarded. That must come later, however, for without a good product all departmental jockeying for advantage is fruitless.

One of the objectives of the task-group relationship is to free the buyer from his identification with Purchasing while retaining his knowledge of purchasing; to free the engineer from his identification with Engineering while preserving his engineering skills; and so on down the line. This arrangement makes it possible for departmental skills to be applied for the welfare of the group as a group. Because the mix of skills parallels the capability of the company or division, the welfare of the group closely parallels the welfare of the company.

And here is where the magic comes in, the magic that always surprises us in a value workshop—the magic of group dynamics. Once a number of men have identified with a group, whether a bullfighting *cuadrilla*, a Bushman hunting party, or a football team on a drive, personal identity is submerged but personal performance is enhanced far beyond individual capabilities.

I know value analysts who can play quarterback in a value analysis team, coaching and playing as they go. They are exceptional people. The ordinary analyst, such as myself, is better off executing the game plan from the sidelines, rather than carrying the ball.

His main task is to get the right people together at the right time, provide them with a meeting place, persuade them to leave their weapons at the door, protect them from interruptions, and dig up the additional information they invariably require.

Riddle of Group Dynamics

We know that group dynamics works, but we do not really know what it is. Behavioral scientists, of course, have observed the advantages of free, many-sided communication, the build-up of creative potential, and the improved personal relationships, all of which are results, not explanations. Gordon L. Lippit (Heyel 1963, page 267) comes near, but no one has really explained the magic that turns a collection of individuals into a source of creativity, productivity, and common sense.

Like many new industrial methods that work well in one area, group dynamics has developed evangelistic followers who believe it will work well in *all* areas.

The same thing happened—early in the game—to value analysis, but because it was applied in the *right* areas it survived the cultist approach, replacing exhortation with demonstration and proving to be a useful tool for profitable competition.

Group dynamics followed the same pattern: search, discovery, and evangelism. Then it began to yield unquestionable benefits from the face-to-face interaction of people working on joint and profitable tasks.

Without pretending to know what group dynamics cannot do, value analysts do know that it is a source of energy and enthusiasm in a value task group.

We did not start out with groups for the sake of groups, much less with the intention of using group dynamics. Much of the early value analysis was practiced by loners, and today most engineers and buyers do some value analysis on that basis, but product value is not contributed by engineers and buyers independent of one another and of the other skills involved.

What happens is that one thread of information about a product is not truly meaningful unless it is viewed in relation to the other threads of the network.

Customer demands without regard for engineering feasibility, procurability of the parts, and producibility of the whole are no more meaningful than engineering specifications based on the assumption that nothing but engineering goes into the product.

To search for, identify, and evaluate the interlaced information required to come up with a good product we have to look at the network *as a net,* not at each thread at a time, but to understand the net we do have to know something about each thread.

Lacking a Leonardo da Vinci, a Leibnitz, or a Benjamin Franklin in the average industrial plant, we have to make do with a collection of specialists, each of whom understands one of the threads and all of whom can discuss the net as a net.

This collection is usually made up of people varying widely in authority, responsibility, education, and experience, but three factors tend to draw them together:

1. Each has information the others need.
2. Each has been assigned a task jointly with the others.
3. Each will benefit personally if the task is successful.

The dynamics of human groups existed long before Dr. Kurt Lewin brilliantly formalized its study, but were it not for Lewin's work I should never have thought of tracing the magic of group dynamics to our ancient hunting heritage nor of using a hunting team to tackle modern industrial problems.

The mammoth, the wooly rhinoceros, the wild bull, and the larger

aurochs were serious problems that troubled our Upper Paleolithic an-
cestors. So were the herds of bison and wild horses whose stampedes could
destroy whole settlements.

How could men, without metal, fight and kill such great beasts? This
is the riddle we are trying to solve.

So we study those fabulous people, but forgetting that magic, science,
and religion were one and the same in those days we categorize their
superb cave art as "hunting magic," putting it all into a cubicle for
"magic."

Let us search a little deeper, however; say we spend a vacation in
France and Spain and actually enter the caves of Lascaux and Altamira.

Once inside the age-old caverns we are overcome by a feeling of religious awe. Our imagination gropes 30,000 years into the past to see the
gifted priests of a lost age painting images on the walls of their cathedrals.
Reverently, we compress the whole scene into a cubicle for "religion."

Cathedrals and shrines these caves may have been, but each was also
the war-room of a hunting community, locked in life-and-death struggle
against giant Pleistocene beasts; and we have no cubicle for *that* category.

What a frustrating thing it is to look at a war-room display and not
be able to attend the briefing! One sees a bison and an ibex wounded by
spears, wild horses and cows in flight, an aurochs charging furiously, and,
at the Dead Man's Shaft in Lascaux, an example of poor tactics—the
earliest record of a bullfighter caught by the bull. A great bull-bison has
thrown a man to the ground. The poor fellow is screaming his head off
for his helpers to attract the bull's attention.

One must certainly agree with Leroi-Gourhan's interpretation of the
caves as shrines or sanctuaries, with the understanding that the transmission of survival techniques and hunting methods is a religious ritual
among primitive peoples.

This brings us to a briefing we *can* attend! Cotlow describes the elephant dance of Congo pygmies on the day after they have hunted and
eaten an elephant. The dance is a detailed rerun of the actual hunt,
emphasizing every important point, from preparation, trailing, detection,
strategy conference, initial advance, withdrawal, reappraisal, and change
of approach to the final attack on a dummy elephant in the middle of
the clearing. Successive spear charges from all sides by the five members
of the hunting party simulate the moment of truth.

Such ritual dances, and similar ceremonies practiced by other modern-
day primitives, reveal that the truly effective aspect of "hunting magic"
is what it does for the hunters themselves, from creating group identity
and commitment to transmitting successful techniques.

We do not have to go to the northeastern Congo or the teak forests of

Burma for a replica of the Upper Paleolithic hunting team. The gods have preserved it for us in the *corrida de toros,* the bullfight.

A more successful bullfighter than the one we met at Lascaux was Gilgamesh, king of Erech, in the land of Sumer. The goddess, Ishtar, made a play for Gilgamesh, and he scorned her. Furious, she invoked Anu, ancient monarch of the North Star. "My father," she prayed, "Gilgamesh insulted me. Make the Bull of Heaven destroy Gilgamesh."

When the Bull of Heaven landed on earth, killing hundreds at every snort, Enkidu, the king's blood-brother and hunting partner, challenged the beast, then leaped aside shouting, "Now! Thrust your sword between the nape and the horns."

Gilgamesh met the charge with a deadly thrust between the nape of the neck and the horns, and the giant bull collapsed at his feet (Sandars).

Now the god responsible for both magic and civilization was Enki, Lord of the Deep, Lord of Wisdom and Friend of Mankind. We can imagine Enki's thundering, "Ole!" as the hunting team of Gilgamesh and Enkidu slew the Bull of Heaven. It was then that he must have decreed: "Hear all ye gods! Lest man lose his hunting magic forever, let us ritualize it into an enduring ceremony, a pageant to be preserved over the generations and carried across the waters."

In far off Crete the Minoan goddess Britomartis-Diktyna must have heard him, for she began presiding over a ritual bullfight, in her topless costume, holding two serpents as though they were *banderillas.* Later, the Emperor Tiberius Claudius ordered the Iberians to cage the fierce bulls of their forests and bring them to Rome, where they performed in the arena. The Iberians used their black cloaks to evade the repeated attacks of the savage beasts before killing them.

The Principles Involved

Behind the success of the primitive hunting team and the industrial task group are age-old feelings of small-group solidarity, feelings that awaken the individual's faith in his own capabilities and in the importance, to the group, of his personal welfare.

The Principle of Direct Motivation

Imagine a manufacturing engineer walking into the factory saying, "Well, Boys, I sold us down the river, but it is for the good of the company."

This would be remote motivation, a leap that neither nature nor value analysis can take with impunity. And why stop at the company? What about the Free World? The solar system? Why not be really noble and say, "Boys, I sold us down the river, but it is for the good of the Universe!"

Nature encourages all living things to survive, grow, and propagate, but it does not skip the immediate goals. The male rabbit does not say to the female rabbit, "Pardon me, but Nature, which is the company we all work for, apparently wants to populate the planets with living organisms. In our case the Grand Plan calls for more rabbits. I will have to load you down with future rabbits for the entire production run—but remember—we are all working for the same company. Be noble!" That is not the way it happens. Nature provides an immediate reward for every step leading to the long-range objective.

If we want our value task group to get off to a good start, we must provide direct incentives for anyone who has to give ground in the interest of the group; otherwise everybody will simply defend his present ideas and there will be no change, no movement, no progress.

Resistance to Change. People usually defend an existing pattern, regardless of its merits. Why? Because change usually involves risk. But is risk that important? Two laws of nature give us the answer: survival and growth. Growth, including propagation, may be the lasting objective, but in point of time survival must come first. The male rabbit of our story cannot contribute to the Grand Plan if he gets killed first. An employee cannot contribute much to the welfare of the company if he was fired the week before.

Personal Risk. Personal risk is usually the first element considered in business decisions. In the analysis of an existing product the principle of responsibility, which we will take up later, recognizes the importance of the risk taker. He has already taken a number of risks in order to come up with a product. With what additional risks does value analysis threaten this responsible person? With the risks arising from sanctimonious hindsight and uninformed criticism.

The decision maker knows that the circumstances that imposed yesterday's actions are seldom known today. He could be helpless in the face of a rehash of what happened in the past and he is naturally wary.

To hurdle this obstacle each member of the value task group must make a conscious effort to understand the reasons for apparently poor decisions, studiously avoiding anything that may cast discredit on the people responsible for the product. The same decent and practical attitude should be adopted toward the other members of the task group. By safeguarding the integrity and professional reputation of the people involved, we release all the energy that would be spent on protective strategy. This energy then goes into the product. First there must be sound, immediate motivation for turning out a really good product. Such motivation must also be based on the laws of survival and growth.

Survival, Aggressiveness, and Territory. When we think of survival, we think of living through floods, earthquakes, tornadoes, and war. The last is probably the greatest menace. Aggressiveness—the expression of a need to fight and win—is certainly a characteristic of the human species. Another is the need for some territory of our own—a defensive base within which we feel safe and comfortable.

Aggressiveness, in a cooperative effort, can be a destructive force. Questioning a fellow worker's professional integrity is a sure way of converting a teammate into an opponent. On the other hand, curbing our own aggressiveness on minor points—such as who had the idea in the first place—can get the job done to everyone's advantage.

The territorial imperative is another matter. We can be as unaggressive as possible—completely pleasant, gentle, and mild—yet if we invade someone else's professional territory we are in trouble. Buyers are often surprised when, in answer to a reasonable suggestion, they are told, "Look, you just buy the stuff! I'll design the equipment." The surprise is no greater than that of the engineer, when he is asked by a buyer, "Are you telling me where to place the business?"

Each has invaded the other's professional territory. Contributions from outside one's domain are desirable and useful provided they are requested; otherwise the other man is telling us how to do our job. To get invited into the other man's territory we have to invite him into ours first.

We have seen that early man's fighting advantage over faster and stronger animals cannot be traced entirely to technology. He could not have developed modern technology if he had been eaten by the lions before he invented the spear. The real advantage lay in the development of the family, the hunting team, and the tribe into effective organizations which controlled individual aggressiveness and broadened the territorial concept. The results of such control were the survival and growth of the individual as part of the group.

We all recognize that survival and growth are basic to life itself, but we often forget that they are basic to many special facets of life, such as earning a living and business success. In business survival means not losing your job, and growth means looking good in the eyes of your boss plus learning something. Every member of the task group must consciously help the other members to satisfy these motives. The resulting commitment to one another's welfare brings about spontaneous recognition of good work.

Once recognition has been granted by a man's co-workers, management recognition comes naturally. A gratifying ratio of dollar gains to man-hours spent results from the direct motivation generated by paying attention to the drives for survival and growth, by redirecting aggressiveness, and by respecting territorial rights.

Members of the value-analysis task group must agree beforehand to get these natural drives working for them rather than against them, to minimize the need for defense, and to maximize recognition of good work.

The Principle of Interacting Skills

When was ever honey made
With one bee in the hive?

Thomas Hood, 1826

Man and the social insects have conquered the surface of the earth by the effective interaction of skills. In our plant there are many people, some bees, lots of ants, but few lions and tigers. The ancient monarchs of the forest have not done too well in competition with social animals. No single human being, bee, or ant has the strength, speed, and cunning of the lion or tiger, but each can develop a specialized skill of its own and use it in concert with its teammates to subjugate stronger and faster creatures.

This interaction of skills has been called *synergetics,* from the Greek word *synergos,* working together. A simpler term is the English *teamwork,* a concept that has suffered from a poor understanding of its true meaning. Everybody senses that teamwork is good, that it is desirable, and that it produces results, but few people bother to learn how it works and why.

Teamwork can be preached until the world itself sounds like a platitude. No amount of preaching, exhortation, or locker-room pep talks can build a team. A team or a task group must be carefully put together to combine the skills required by the nature of the planned task. The team members must be trained to apply these skills in patterns that have proved successful in actual experience. That is how all "rules of the game" and "combinations of plays" are developed. The men and women who make up a team must be committed to the attainment of a common goal and must be willing to obey the signals of a quarterback.

The English word *teamwork,* if we strip it of empty preaching, means more than simply working together. It implies the unstinted recognition and use of the special skills of other team members, commitment to a common goal, and an understanding of the playing field or, in our case, of the industrial environment.

The British Navy, during the Napoleonic wars, offers a fine example of effective teamwork. Nelson met and defeated every fleet a united Europe could send against him. He established such working relationships among wind, sea, ships, and men that even the enemy played a directed role. Fulfillment of this man's genius was made possible by the discipline, skill, and composition of every ship's company in the fleet.

The teamwork of value analysis is based on the composition of the team which must include engineering, purchasing, manufacturing, cost estimating, and marketing skills whenever possible. Closely related skills may be substituted as long as the technical, socioeconomic, and financial aspects of product value are covered.

To make a balanced combination of skills really effective we have to face up to the existence of professional and developmental interests. The characteristic outlook of the engineer and of the engineering department is quite different from that of the buyer and the purchasing department. This is true of Finance and Marketing, as well. The interests of the departments vary in the same way.

When two of these interests clash, there is usually some personal conflict, but when all pertinent interests are represented there is a plurality of emotionally detached people who serve as arbiters to compose differences in the interest of the whole. Such contrived objectivity puts the joint wisdom and sense of justice of 12 ordinary men on a par with that of the judge on the bench. The judge contributes knowledge of the law and of courtroom procedures; the jury contributes objectivity and impartiality.

In business, as in the Congress and the courts, we have to provide a mechanism for the kind of group dynamics that achieves objectivity.

The Principle of Objectivity

Impartiality, detachment, and objectivity all refer to the appraisal of information on its own merits, as distinguished from what we would like to see or would like to hear. If the facts are unpleasant, we may simply say, "That is not true!" This trait, of course, is more noticeable in others than in ourselves.

Self-deception is, perhaps, the greatest obstacle to objective thinking. It is not a newly discovered trait. In Book III, Chapter 18, of the Gallic Wars Caesar writes that he defeated the greatly superior forces of Viridovix by deception, "because men generally believe willingly what they want to believe." In his third *Olynthiac* Demosthenes warned, "Nothing is so easy as to deceive one's self." Sixty-four years later another Demosthenes, the general, set out with a huge armament of Athenian ships and men to conquer Syracuse.

The home team, inferior in numbers, but fighting in their own harbor and on their own land, completely crushed the expedition and executed its leader. "When the news reached Athens," writes Thucydides, "they for long refused to believe that the armament could have been so utterly destroyed."

Sometimes we feel like shouting, "Get your head out of the sand!" This is really an unreasonable request. Keeping the head in the sand generates a certain measure of tranquility which makes an ostrich's last moments happy until the hunter reaches him. Who are we to say to the ostrich, "You are sacrificing long-range goals for the sake of feeling good now. Stop this self-deception. Look at the facts calmly and objectively!"

The truth is that we cannot redesign people or ostriches to be more objective, but we can combine them so that biases balance-out, thus generating a greater measure of group objectivity. For this reason it is customary to include in the value group representatives of the required skills who are not emotionally involved with the product. The result is a broad look at the consequences of value analysis in terms of group welfare as well as individual welfare, in terms of company gain as well as departmental advantage.

The Principle of Effective Communication

Get two design engineers, a cost estimator, a buyer, and a manufacturing engineer together and the conversation will flow easily about such matters as cream and sugar in the coffee, the recent ball game, and crabgrass. Then the electrical engineer explains why he cannot change a special part for a standard one. Pointing to a chart on the table, he says, "You can see that even a slight change will depress this parameter by a whole order of magnitude. Besides, we have to have the best Q we can get."

"The best Q?" asks the buyer. "That brings to mind what I called this meeting for. If you specify a 'special,' we will drop below the EOQ."

There is a moment of bewildered silence; then the Manufacturing engineer tries to clarify the matter, "EOQ is Purchasing's version of our ELS. And we are in the same boat. If you don't use a standard item, *we* go below the ELS. PC will have to make two pulls instead of one. We may have to restationize and even *refacilitate!*"

Overawed, the design engineers look helplessly at the chart, then at each other. "Perhaps," suggests the electrical man, "we could—"

The mechanical engineer is shaking his head. "I am afraid not," he declares, spreading out another chart. "It would take us beyond the bounds of permissible parts density—actually beyond the asymptotic line of the parts density curve."

Of course, if it does *that*, the whole project is abandoned. These people may as well not have met at all, but this hypothetical meeting does illustrate how the secondary functions of language can nullify the primary function as a means of general communication.

One of these secondary functions is to present the social and educational background of the speaker, as illustrated in Shaw's *Pygmalion* and the theatrical hit, *My Fair Lady*. A closely related one is to establish the speaker as a knowledgeable member of a craft or profession. A third and more important function is that of specialized communication within a craft or profession. This can become a habit—a costly habit when general communication is important.

An early task of the value-analysis workshop is to achieve effective communication by encouraging all specialists to use as much plain English as possible. For this reason the value analyst himself must learn and practice the art of plain talk.

The Art of Listening. A sure way to lose information is to interrupt a speaker and question his premises before he arrives at the conclusion. By doing so we immediately reverse the flow of information. We *give* him the information that his logic is faulty or his facts are wrong but we *lose* the information he was making available to us. A classical example is the story of the ambassador who was kicked over the Army goal posts in the Philadelphia stadium. His chauffeur tried to warn him, but instead of listening, the ambassador growled, "You fool, there are no mules on football fields. You are thinking of polo, and polo is played on . . . ."

He chose to give information rather than receive it by telling his chauffeur that he was a fool, that, as a rule, there are no mules on football fields, and that polo is played on horses, not mules. But he lost the useful information that he was backing into the business end of West Point's mascot, the Army Mule.

Similarly, in industry many people are kicked over the goal posts into less rewarding jobs because they never got the message—they interrupted to show how much they knew. On the other side of the goal posts these brilliant folks find others whose specialty is the question that hinders.

The Art of Inquiry. If what we want is information, our questions should not give the speaker a hard time. They should help him make *his* point, in *his own way* and at his own rate of progress. If listening is to our advantage, we should work just as hard at understanding as we do sometimes at not understanding.

A mother training a child or a theologian converting a heretic may use questions to prove or refute a point. Such questions are answered with the wariness of an animal skirting a trap or with the unwillingness of a bull being led by a nose ring. The child or the heretic may acknowledge his errors and agree to change, but the questioner has gained little information.

There are directed questions, however, which help the speaker make his meaning clear. The best example of such helpful inquiry is provided by a good trial lawyer when examining his own witness.

Care and Nurture of New Ideas

Like any other infant a new idea coming into the world is frail, delicate, and frightens easily. It has to sneak into the mind of its host, elbowing its way past many established concepts to gain the attention and recognition that keeps it alive. If the new idea is lucky enough to arrive at a time when the host has time to think about it, the host might even talk about it—share his new idea with someone else.

That is usually the end of the new idea. The other person will say, "It won't work!" And he will know *why* it won't work. Not just one reason. He will know *all* the reasons why it won't work.

For years it was believed that the shortest interval of time known to man was the interval between the time when a traffic light turns green and the time when the driver in back blows his horn. It turns out that there is a much shorter interval of time—the time it takes the average person to know why somebody else's idea will not work.

If we want to profit from the ideas of others, we must welcome the infant ideas into the world, encourage them, coo over them, and give them a chance to grow. With a little help they can develop enough strength and confidence to show their good points.

Not only must we protect and nourish infant ideas, but we must also provide them with playmates—with other new ideas. Once we have a large selection we will have a fair chance of finding good ones. If we slap down the other man's idea before it has had a chance to develop, we are the losers, for our own ideas need playmates to develop their full potential.

Later on we will have a formal innovation session for the express purpose of generating a large selection of new ideas, but some are bound to come up before that time. Let's give them a chance.

Following the innovation session—when the young ideas are strong enough to face the world—we can look at them critically. In the meantime let us solemnly agree not to say, "It won't work," about anybody's new idea.

Task Group Effectiveness: A Summary

A major advantage of the task group is the mixture of suitable skills. The diverse specialists, communicating face-to-face, can see at once the effect of their actions on the activities of others. They go to work on improving the product in a way that reflects credit on each member of the group and on the persons who have entrusted their product to them. This is a matter of cold business logic. Entirely too many company resources are wasted in building defenses against unwarranted criticism. The simplest way to reduce this defensiveness is to eliminate the attack. All the effort that previously went into sharpening knives, procuring Band-Aids, writing "memos to file," and covering the body's vulnerable opening now goes into improving the product.

To get the most out of their interaction team members have by now learned to achieve a measure of objectivity, to communicate without jargon, to listen cooperatively, to inquire productively, and to work well together. This makes for a powerful tool indeed. Now for the way to use this tool.

WORKSHOPS AND SEMINARS

When a number of task groups work together on live products, they constitute a workshop, as distinguished from a seminar. A seminar is primarily a learning experience, whereas the immediate function of the workshop is to improve the company's gross margin between sales and cost. Schedules for training seminars are fairly cut and dried, but the distribution of time among the phases and tasks of a workshop depends on (a) the nature of the projects, (b) the type of workshop or seminar, (c) time available, and (d) feasibility of the schedule.

Nature of the Projects

"No matter how much you would like to value-analyze this or that area," my boss once told me, "you must first value-analyze your own operation, and that will inevitably lead you to dig where the pay dirt is."

The richest lode of the pay dirt and the fattest nuggets in that lode are the following:

Designs which have to advance the state of the art. The fiercest competition today is in the race for new products—new ways of doing things. For this leap into the unknown an engineer needs, more than ever, fresh information on new materials, new suppliers, and new manufacturing methods.

Products which must be delivered ahead of schedule—getting there first. Where there is no time for the step-by-step approach characteristic of yesterday's gentler competition, a value task group can do concurrently much that was done sequentially.

Jobs that cost more than they should, either because the price would exceed what the customer can pay or because the gross margin does not yield enough profit.

Workshops and Seminars

To gain a greater return from his efforts the experienced analyst coaches more than one group at a time, usually 3 to 10 groups or teams, although an important new product may justify a single blue-ribbon task group. When there is a pool of people who have attended good training seminars or who have participated in several task groups, they can be organized to look at a product without benefit of a value specialist.

Most value analysis, of both present and future products, is conducted in workshops of some two to four dozen people, backed by a supporting group to provide information and prepare for implementation. How the plant training manager and the value analyst handle the details of workshops and seminars is explained in detail in Chapter 10.

The Support Group

This group varies from a list of individuals with whom members of an extended workshop must touch base and develop more information to a formal group, gathered near the area of a quick-action, intensive workshop. This group makes itself available for consultation and is ready to provide or to dig up additional facts. In either case the designer or product manager is personally represented in the support group.

Total Elapsed Time Available

Lots of time, like lots of information, can do more harm than good. The task group works best under pressure. The results of too much time are (a) loss of momentum and (b) a tendency to do or redo the work of the line specialists instead of improving the effectiveness of their interaction. Too little time results in frustration and ineffectiveness. Workshop duration for the great majority of products ranges between a minimum of 32 hours, an optimum of 48 hours, and a maximum of 96 hours.

A Feasible Workshop Schedule

Meeting half-days, Tuesdays and Thursdays, for two weeks constitutes a 32-hour-plus workshop. The plus is the time "made" during the rest of the week for gathering additional information. In a small plant this may be the only way to get the people away from their regular duties.

The full-week, eight-hour-day workshop has excellent momentum, generates great enthusiasm, but allows little time for getting new information, making and testing prototypes, and receiving samples from vendors.

The same 40 hours can be spread over two weeks to provide the elapsed time needed, but as long as you are spreading it out a 48-hour workshop, meeting all day Tuesdays, Wednesdays, and Thursdays, is just about right for that period. Add one week and three more workshop days and you have a luxuriant 72-hour workshop for very special new products.

THE VALUE-ANALYSIS JOB PLAN

The task of analyzing value follows the general pattern of the scientific method, incorporating problem solving and innovation techniques with the teamwork characteristic of group dynamics.

We owe the original and very successful sequence to Larry Miles, who was using it in 1947. Anthony R. Tocco presents an excellent version of it in the *Encyclopedia of Management* (Heyel, page 1026).

The three phases, *information, analytic,* and *creative,* by whatever name is given them, are the core of the job plan; all the rest is prologue or aftermath (Table 4-1).

Table 4-1 The Job Plan

Preparation
Information phase
Analytic phase
Creative phase
Evaluation
Presentation
Implementation

Preparation

This task starts two months to two weeks before the workshop. It includes selling the workshop to local management, selecting projects and participants, scheduling the effort, requesting the facilities, and preparing a data package for each project. Such a package usually contains

marketing requirements
engineering specifications
costed-out list of materials
drawings and schematics
manufacturing costs
quantities to be produced per year
manufacturing schedule
manufacturing process
contract information
data on applicable . . .
   new materials
   new processes
   new products
   new suppliers
company standards
industry standards
government standards

As soon as the participants have been selected, the plant training manager sends them preliminary written material. Ideally, the material should be sent to arrive at the person's *home* as follows:

A month in advance: *Function of Value Analysis* and *Nature of Value,* both condensed.

Three weeks in advance: *The Principle of Consumer Sovereignty* and *The Principle of Concentration.*

Two weeks in advance: *The Principle of Direct Motivation* and *The Principle of Interacting Skills.*

One week in advance: *The Principle of Objectivity* and *The Principle of Effective Communication.*

So much for preparation *before* the participants arrive. Once they arrive, a senior member of management welcomes them, and the training manager organizes the task groups, assigns the projects, and the workshop goes on to . . .

The Information Phase (Chapter 5)

Before embarking on the search and collection of data the task groups
study information itself: its nature, usefulness, and how to identify and
select the portion that is worthwhile.

Principles applied in this phase:
   adequate information
   selectivity
Main task: investigation
   identify the project
   define the scope of the study
   determine quantities and product life
   learn marketing requirements
   review cost data
   consult company specialists
   consult suppliers

The Analytic Phase (Chapter 6)

This phase was developed to answer the question, "What does it do?"
It is concerned with *the function* of the product.

Principles:
   direction
   responsibility
   usefulness
   limited resources
   economy
Main task: analyze the function
   define task group goals
   relate goals to company objectives
   define product functions
   determine actual or estimated cost of
     providing each function
   determine minimum or basic cost for
     which each function can be provided

The Creative Phase (Chapter 7)

For a man on foot to kill a wild bull with a sword, as Gilgamesh did in
ancient Sumer, is a singular accomplishment. Of course, Gilgamesh was

a nine-foot giant, and he had Enkidu to help him. It turns out, however, that with a *cuadrilla* of five helpers an ordinary man can kill a wild bull with a sword, but each of these helpers is a specialist in his own right.

In the same way a value task group, made up of ordinary people in the right combination, can be productive far beyond the sum of individual capabilities.

Principle: balance and proportion
First task: innovation
   create climate for new ideas
   surmount barriers to innovation
   generate new ideas
   screen ideas developed
Second task: simplification
   elimination
   combination
   rearrangement
   modification
   substitution
   standardization

Evaluation (Chapter 8)

Having developed a number of ideas and removed the scaffolding of the creative process, we must now test them out by making actual models, if time permits, compare them, and recommend the viable choices for selection by the decision maker. There are three steps:

Verification
Comparison
Choice

Presentation (Chapter 9)

Presenting information to management should be one of the most condensed and honest forms of communication. Condensed because management already knows a lot about the business and needs only the missing parts of the jigsaw puzzle. Deliberately honest because the tendency, when speaking before the Councils of the Mighty, is to present a good case. The function of a value-analysis presentation is not to persuade or plead but to inform. This function is as simple as one, two, three:

Preparing the report
Preparing the visual aids
Making the presentation

Implementation (Chapter 9)

The results of a value-analysis effort are not results until they have been implemented. Opening the door to implementation is all that the presentation to management does. The analyst, the value service group—if there is one in the company—and often the team members themselves have to walk through that door carrying a rich burden of information and plans. All the interesting details, minor problems, and minor opportunities that we threw out of our initial presentation must now reach the people who will have to deal with them. There is someone, somewhere, who will have to change that winding cord on the new line of talking dolls. All we found out about the cords, our recommendations, and the product manager's decision must be presented to the foreman of voice-box-assembly, and so on for every recommendation. Among the steps listed below the first one is the most important.

Presenting the information
Forecasting the consequences
Finding a sponsor
Obtaining
   authorization
   skills
   facilities
   resources
Scheduling
Follow-up
Reporting results

THE MOST DYNAMIC OF GROUP DYNAMICS

My favorite illustration of how a task group works is the Spanish bullfight or *corrida de toros* which we have followed so far from Upper Paleolithic times to the present day.

I will ask the reader to accompany me to the Plaza de Toros in Malaga on a Sunday afternoon and to learn a little Spanish on the way because I cannot bring myself to call a *toro bravo* just a bull!

The Project

El Señor Toro is a wild beast of the forest whose forbears never pulled a plow nor saw the inside of a barn. He is our value analysis project for this sunny afternoon.

We walk briskly up the hill from Malaga's pleasant harbor to the Plaza de Toros and take possession of our choice seats.

Preparation

The banda taurina strikes up the majestic first bars of *La Virgen de la Macarena* as the toreros, the bullfighters, march in, easily and confidently, to stand proudly before the presidencia.

At Knossos, in Crete, the presidente would have been a young, beautiful, and topless votary of the Sweet Virgin Britomartis. In Rome it could well have been the Emperor. In Spain it is now *la autoridad*, a nice flexible term.

After bowing to the presidente, the toreros on foot file out past our point of vantage, leaving the arena through the burladero de los matadores, one of the four escape openings in the barrera around the ring. The picadores ride out of the ring through the puerta de caballos.

Cuadrillas

La cuadrilla de toreros is the matador's team, his task group. The trumpet call which starts this phase is called toque de cuadrillas and its message is "task group, you are on!" It is addressed to the cuadrilla of the first matador. One of the junior peones de capa—literally footmen with capes —steps out from a burladero not far from where el toro is to appear. Suddenly the door of the toril slams open and el toro storms out, snorting and full of fight. He sees the man in motion and charges the burladero, knocking the shield to splinters. From *our* burladero the senior peon springs out and begins caping the beast with wide-open lances, running him back and forth before our burladero, where the matador is watching.

"All right, all right," says the matador, "you've shown me his right horn twice. I know he hooks to the right. If you do that again, you'll be talking in soprano! Now, show me his left horn . . . ."

Information Phase. The peon de capa is eliciting information on the behavior of the bull, for this is the information phase of the value-analysis job plan. The matador himself now capes el toro, using veronicas to pass him close to his own body, to learn the beast's behavior at close quarters.

When the presidente sees that the matador is satisfied, he signals for the kettle drums and trumpet to call out the . . .

Picadores

These horsemen, on padded horses and armored legs, receive the charge of el toro on the point of their long picas, so that the animal hits horse, man, and pica at the same time, thus reducing his excess power.

Cutting the Problem Down to Size. Not only do the picadores absorb much of beasts energy but they physically cut him down to size by making it painful for him to raise his head too high. This encounter is the first step to make el toro forget that the cloth is only a lure.

Banderillas

When el toro has received three or four picas, the kettle drums call for silence and the trumpet sounds el toque de banderillas. *Bandera* means *flag, banderilla,* means *little flag,* and *banderillero* means the man who sticks banderillas into toros.

What for?

El toro is not only getting suspicious, he is beginning to think—a very bad trait in an enemy. To allay his suspicion the banderillero must come near without any waving cloths and he must show el toro that this thing that he is fighting is quite real and can hurt him.

From a good distance away the banderillero holds the sticks high, challenging el toro, and as the beast charges so does he. Another banderillero pops up to one side, el toro turns toward him, and whamo! . . . he gets it.

Our project is back on the track. After three pairs of banderillas el toro has learned that he must concentrate on the nearest *moving* object and forget the rest. Also, he is so angry that he has stopped thinking.

Implementation

The booming of the kettle drums lasts a little longer this time and the trumpet's toque de muerte, or death call, ends in a drawn out, foreboding note. Now the matador is left alone in the arena. Making whatever passes are necessary, he now squares off el toro, working him until the animal's forefeet are close together, forcing the shoulder blades apart for the sword thrust. He raises the estoque, sights along the shining blade, and sprints forward. Going over the horns, he delivers the death thrust in the moment of truth.

Who *is* this man who can kill a wild bull with a sword?

Lessons from the Corrida de Toros

Is the matador a descendant of the Spanish nobility who have been practicing the military arts for generations or perhaps a privileged youth of the industrial class whose parents spent a fortune in his training?

Not at all. Most toreros are ordinary people. Many are children of the poor and underprivileged. The virtue of a task group is that it produces exceptionally good results, working with people as they are.

Human potential is without measure. There is no limit to what ordinary people can do when they combine skills to support each other in pursuit of a common goal.

Lessons that any task group can learn from a matador and his cuadrilla are these:

Each class of torero has a special skill that the others need.
No one skill functions well without the others.
The whole group benefits from a successful performance.
Each man depends on the others for his welfare.
They stick the picas, the banderillas, and the estoque into el toro and not into each other.

V

INFORMATION PHASE

Finding buried treasure is as much a matter of information as it is of treasure. Consider the delightful adventures of those professional treasure hunters who search the bottom of the sea.

Competing with many well-financed and lavishly equipped search vessels, Robert Stenuit, a Belgian professional diver, and Marc Jasinski, his photographer and friend, discovered and salvaged a dream-haul of treasure from the Spanish Armada (*National Geographic*, June 1969). Their equipment? A rubber raft, an outboard motor, scuba diving gear, and a vast store of scientifically gathered information. These young men had spent years in the archives and libraries of Europe studying every letter, every report, every legend that referred to the Spanish ships *Duquesa Santa Ana, Sancta Maria Encoronada*, and the galleass *Girona*. How they found the treasure of all three ships *in one spot* is a fine example of the value of information.

Beauty, Truth, and Information

The turquoise-blue sea between Florida and the Bahamas is particularly beautiful in January and February. Its white caps progress gently, in orderly array, under a clear blue sky. At this season, far to the north, the weather is ugly. Skies are overcast and sullen; dark gray swells conceal dangerous shoals. The beautiful Caribbean, on the other hand, tells us the truth, shows us the shoals in light green and the clear channel in deep blue.

If beauty is sometimes an indication of truth, truth itself is a measure of the quality of information. The first step in the analysis of value is the analysis of information.

TO BE INFORMATION IT MUST BE NEW

The beautifully useful communication theory developed by Nyquist, Hartley, Kolmogoroff, Wiener, and others, and culminating in Shannon and Weaver's *Mathematical Theory of Communication*, has come to be called *"information" theory*, which suggests a general theory of information; but that is not what it is at all. Shannon named it correctly; it is a theory of communication, not of information in general. His mathematical conclusions on the transmission of information, however, do help us to understand information itself, but "in a sense, it is a pity," writes Cherry (1966), "that the mathematical concepts stemming from Hartley have been called information at all." What concerns us here is the *statistical rarity* or *surprise value*, for that is an area in which communications theory can enlighten us on the very nature of information.

Wilson and Wilson put it this way:

"Unexpectedness. 'I got little information because I knew what he'd say before he opened his mouth.' . . . Stated another way, if the received message is predictable, no information is gained by receiving it."

Pierce makes the same point, which is indeed one of Shannon's major building blocks—the concept of inverse probability in the transmission of signals. How inverse probability enters the concept of information in the everyday sense of the word is best illustrated by the question, "So what else is new?"

Xenophon tells us that Cyrus the Great would instruct his scouts to "send him reports from time to time of whatever they saw that was new." To be news information must be *new*. Much supposed information is simply the reaffirmation of comforting old facts. This veil of platitudes hides real information.

IS IT TRUE?

Here we must depart from communication theory, for false information can be communicated just as readily as true information, and false information increases ignorance instead of knowledge.

Since communication theory is primarily concerned with the transmission of messages, it ignores the veracity of the source. Wilson and Wilson assume that all messages generated by "a source" are error free. The assumption is valid in the theory of encoding, transmitting, and decoding messages, but it cannot be applied, nor do Wilson and Wilson intend it to be applied, to information in general.

Imagine a master spy ignoring the judgment, accuracy, and truthfulness of his sources of information!

Why talk about spies?

Because spies are professionals in the field of information. So are researchers, but conflict being the potent motivator that it is, mankind to date has given much more time and effort to planning, organizing, and directing espionage than it has to planning, organizing, and directing research. Here is an early example:

When Saul instructs his spies to locate David (1 Sam. 23:22-23; 1000 B.C.) he does not ask for a bundle of uncertainty; instead, he directs:

> . . . see his place where his haunt is, and who hath seen him there . . .
> and take knowledge of all the lurking places where he hideth himself,
> and come ye again to me with the certainty (King James Version) with
> sure information (Revised Standard Version).

Saul's request, of course, implies the newness of the information. He would not ask "where his haunt is" if he already knew it, but what he emphasizes is certain or sure information. He not only wants the unknown revealed, he wants it revealed accurately and with reasonable precision.

In saying that information should be true as well as new, I do not mean to disparage the advances of communications theory. Starting out with newness as an accepted property of information, communications theory has developed an accurate and precise measure of the amount of information that can be transmitted over a given channel. It shows how to distinguish between the possible and the impossible.

SOME DEFINITIONS PERTAINING TO INFORMATION

VIGILANCE. Vigilance is a state of general watchfulness and keen receptivity to information. In the days of pirates and privateers the first person to sight a prize was rewarded with a larger share of the loot for his vigilance. A lookout at sea, an impoverished father standing guard to make sure nothing interferes with the elopement of his daughter, and the outside man in a gang of burglars, all exercise vigilance.

ALERTNESS. Alertness denotes readiness for action, as well as sharp attention. More dynamic than simple vigilance, alertness utilizes existing information to focus attention on selected items of new information. One is alerted to meet an infantry attack or to catch the fish while they are running. A scanning radar is vigilant, but a tracking radar, ready to lock on the target, is alert.

RAW INFORMATION. Raw information is the material that spies and intelligence observers collect. It is a special case of information-in-general, as defined below. Raw information is made up of bits and pieces—loose facts, so to speak—and of information on the sources of such bits and pieces.

INFORMATION-IN-GENERAL. This category is the building material of knowledge; only *new* knowledge can be added to old knowledge to yield more knowledge. In this sense information-in-general and the "information" of communications theory agree. Information must be new, but that is not enough; it must also be true or nearly true. How closely information approaches the truth is measured in two ways:

accuracy, which is freedom from error, and

precision, which is the exactness or degree of refinement with which measurements are made or events described.

INTELLIGENCE. Intelligence is the name given to raw information after it has been evaluated for pertinence, credibility, and accuracy, in the light of other information available.

Intelligence (G-2)

Intelligence is an age-old military and political activity. King David, on learning that his rebellious son Absalon is marching on Jerusalem, abandons the city, but not until he has left behind him an intelligence network complete with a master spy, a center of communications, two safe houses, and a system of couriers, protected by a cutout—a person who could be removed to break the chain.

He instructed his friend Hushai to receive Absalon as King and to say to him, "I will be thy servant, O King; as I have been thy father's servant hitherto." Then David added, "And hast not there with thee Zadok and Abiathar the priests? Behold, they have there with them their two sons . . . and by them ye shall send unto me everything that ye can hear."

One of the safe houses was just outside the city walls near the spring Enrogel. There the young men, Ahimaaz and Jonathan, hid in safety while the cutout—one of history's truly anonymous spies, called only "a wench"—brought information from Zadok and Abiathar.

The second safe house was at Buharin on the road to David's camp. This house had a well which was easily concealed and in which the couriers could and did hide.

About 400 B.C. Ho Yen-hsi, in his commentary on Sun Tzu's *Art of War* (Griffith), mentions that a Director of National Espionage appears among the titles of military officers listed in the Book of Rites of the classical Chou period.

Here is what Sun Tzu says about employment of secret agents (after Griffith and Kuo Hua-Jo):

> There are five kinds of secret agents: native, inside, double, doomed, and live. . . . *Native agents* are natives of the enemy's country, who are in our service; *inside agents* are enemy officials in our service; *double agents* are enemy spies in our service; *doomed agents* are planted to give the enemy false information; *live agents* return with information. . . . Basics in dealing with enemy spies are: find them, bribe them, treat them well, and instruct them; then turn them around as double agents to recruit—for us—native and inside agents in their own country.

In 1653 John Thurloe, Oliver Cromwell's secretary of state, was given charge of the "department of intelligence." Thurloe's own personal secretary, however, was an agent of King Charles II, whom Cromwell had driven into exile. When the King returned, he knighted his agent, ". . . and the King did give the reason openly, that it was for his giving him intelligence all the time he was clerk to Secretary Thurloe" (Dulles).

Thus the word *intelligence* enters official language as the craft of finding out what enemies and potential enemies are up to. This activity was handled in the United States, at the turn of this century, by the Military Information Division of the Adjutant General's Office. In 1903 the division was incorporated into the Army General Staff as the "Second Division," hence the designation "G-2." Should the reader at this time choose to forget value analysis in order to become a master spy, he is urged to read *The Craft of Intelligence* by Allan Dulles. He will not only enjoy a fascinating evening—you can't put the book down—but he will become all the more interested in value analysis itself.

The task of collecting intelligence (G-2) includes espionage, reconnaissance, and exploration, all of which call for vigilance, alertness, and systematic search.

Espionage

You do not have to look around for another James Bond; *you* may be the one! Espionage is all around us. Here are some examples: (a) a design draftsman sneaks into the factory to find out how they take out the tool after forming a part; (b) while his drawing board is unguarded, a buyer is furtively listing the purchased items on the drawing in order to gain lead time; and (c) while the buyer's desk is unguarded, a manufacturing engineer is hurriedly shuffling through recent purchase orders to find out what the machine he ordered really cost.

In a value-analysis task group these specialists or their bosses make a treaty of peace and the espionage is reduced to voluntary exchange of information, with considerable saving in elapsed time.

Reconnaissance

Going after specific facts that you know or believe are there is the distinguishing feature of reconnaissance. "Spy out the land of Canaan," Moses told his reconnoitering party. "See the land, what it is; and the people that dwelleth therein, whether they be strong or weak, few or many." Reconnaissance seeks information of known value. But do we always know what information will be of value? Many of the most useful finds in aerial photo-intelligence turn out to be something different from what the mission went out to photograph.

To create such information windfalls on purpose we can resort to the technique of . . .

Exploration

When the bear went over the mountain to see what he could see, he was exploring. He did not have to know what he was looking for. The most successful forms of life, such as social insects, mammals, and birds, explore deliberately and systematically. This effort is not the patterned search for food, a mate, or a nesting site. It is a random search for apparently unnecessary and possibly useless information. The only reward is information for the sake of information or, to put it another way, exploring for the sheer fun of exploring, where fun is the incentive to gather apparently useless information. Such "useless" information may prove vital in a sudden change of circumstances or it may prove surprisingly useful following a moment of insight—a revealing change in the way we look at what we already know. More important, unexpected, unsought information often triggers a flash of insight.

FROM SEARCH TO INNOVATION

Scores of well-conducted, well-documented experiments, both in education and in industry, reveal that the most creative individuals in a given group are the most thorough in the information search and show a propensity to explore more widely and actively than the others. Examples of such studies are in Karlins, Lee, and Schroeder.

The kind of learning that accompanies exploration not only provides grist for the mill of creativity but also contributes to the development of intelligence in the broadest sense of the word.

Before launching a value task group on the quest for information it is the usual practice to tell them, "Don't leapfrog. While you are on the Information Phase *stay* on it. Don't invent anything now because it may not be applicable later. Don't have any ideas."

Of course, all sorts of ideas immediately pop into the heads of the participants. What to do? Squelch the ideas? We may forget them. Talk about them? No—they will get clobbered. Write them down. Save them for the Creative Phase.

COLLECTING INFORMATION

The Search for Eldorado

In the days of Don Carlos V, King of Spain and Holy Roman Emperor, just before Christmastide of the year 1535, Don Gonzalo Ximenes de Quesada, chief justice of the newly established colony of Santa Marta, in what is now Colombia, began weaving together all the threads of his information net—Sun Tzu's "divine skein" (Griffith, p. 145).

Each thread told a different story, creating a fanciful vision—a green thread for tales of emeralds, blue for reports of distant mountains, yellow for the legend of El Dorado—the Golden King of the Chibchas—and all of them intertwined with a beige warp, representing the Great River, flowing majestically from the interior.

By Lent of 1536 Don Gonzalo had enough information to set out in search of El Dorado. He led 500 men along the river bank and 300 more followed in flat-bottomed brigantines.

Lessons from Successful Searches

"We found gold coins minted in Seville carrying the crown of Aragon," writes Stenuit. "A sweep of my right hand exposed a yellow object . . . a gold medallion." And he goes on to describe the finds of a very successful search.

Carlos Quinto (Charles V), Holy Roman Emperor and King of Spain, characterized another search as successful:

FORASMUCH AS THOU, GONZALO XIMENEZ DE QUESADA, didst go to our Indies of the Ocean Sea, and following the Great River, in the midst of warlike Indians, thou didst conquer and pacify the natives of the land, putting all under our Royal Yoke and Lordship, all of which to serve us as a good and loyal vassal: WE COMMAND THAT THOU BE GIVEN FOR ARMS: A shield in two parts; in the uppermost, a golden lion holding a naked sword in a field of red, in memory of thy spirit and fortitude in going up the Great River to discover and win the New Kingdom, and in the lower half, a mountain arising out of the sea and covered with emeralds, in memory of the emerald mines thou didst discover. Given in Madrid, this day, the twenty-first of May, in the year one thousand five hundred and forty-six.

I, the King

What can we learn from these successful searches? The first and least expected lesson is that the search for information often takes courage— courage in proportion to the expected worth of the prize. Whatever else may be said of spies, military scouts, and explorers, they are not cowards.

The bloodcurdling passages in Robert Louis Stevenson's *Treasure Island* all hinge on the search for information. The book illustrates what Gabor describes as the value of exclusiveness in information.

The line between magic and science is still indistinct when it comes to exclusive knowledge. The less sure a specialist is of his facts, the more he resents others prying into them, not because he is bad but because he is people.

Courage, therefore, is required to ferret out information, to ask unwelcome questions, to point out errors in statistical calculations, and to remedy goofs and boo-boos that could explode into major flaps.

Courage alone seldom wins victories, as Ares, ruthless God of Battles, must have found out to his discomfiture. He was always getting clobbered while Athena—rational goddess of prudent and intelligent tactics— smiled placidly and picked up the marbles. Even better than Athena's tactics are those of Sun Tzu, "The acme of skill is to win without fighting" (Kuo Hua-Jo), or—in our terms—to get hoarded information without offending the hoarder. Such a task calls for resourcefulness and tact as well as courage.

When a clerk says, "Sorry, we are not allowed to give out that information," his boss may be easier to deal with. He may well say, "We'll be glad to help you. Why do you want the information?"

Your answer may decide the difference between success and failure.

You can say, "I have a right to it and I demand to see it," to which your opponent—now he *is* an opponent—will retort, "You'll need a search warrant and a subpoena, and by the time you get them the information won't be here."

Or you can level with him. Tell him exactly how you intend to use the information, making it clear that you will protect his interests in return for his help.

The second lesson tells us that successful searchers define and carefully identify their objectives.

Robert Stenuit, the diver, was searching for treasure; not any treasure, but underwater treasure; not any underwater treasure, but treasure lying in relatively shallow coastal waters; not any coastal treasure, but treasure of the Spanish Armada; not the treasure of any of the Armada's vessels, but the treasure of the ships *Duquesa Santa Ana,* and *Sancta Maria Encoronada,* accumulated aboard the galleass *Girona.* He carefully identified what he was looking for.

Don Gonzalo Ximenez de Quesada was looking for El Dorado, all right, but to him El Dorado meant not only a Golden King but a kingdom. After all, you cannot eat gold, wear it, live in it, or make love to it; but a well-organized Indian community could become a new kingdom, as he said when he took possession:

Before God, and in the name of Don Carlos, the king my master, I take possession of this land, and I call it the New Kingdom of Granada; and on this site I now found a city, which in honor of Santa Fe de Granada, I name Santa Fe de Bogotá. Done this sixth day of the month of August in the year one thousand five hundred and thirty-eight.

If the legend of the Golden King lent glamor to the search for El Dorado, the overriding theme was resources—land, people, corn, salt, and potatoes, as well as gold and emeralds. Don Gonzalo knew what he wanted.

The search was successful in that it found the makings of a nation—a nation whose original inhabitants accepted not only Don Gonzalo's rule but also the farm implements he brought with him and the culture of sheep and cattle.

LEARNING MARKETING REQUIREMENTS

Simply knowing what the customer really wants today can improve a company's competitive position, a buyer's effectiveness, an inventor's bread and butter, and a housewife's pleasure over a lovingly prepared dinner.

Let us say that the little wife buys new candles to set in the silver candlesticks, flowers for a centerpiece, a bottle of Chateau Haut Brion 1963, and two tender Cornish rock hens to serve in nests of wild rice—

all in loving preparation to celebrate the event that her lord and master has been rated *expert* by the American Motorcycle Association.

He comes home to make the unforgivable statement, "That's what I had for lunch."

The idiot! What does it matter? The little wife had prepared not 600 calories of nourishment but a love rite, and the dolt was thinking only of eating midget chickens!

The interesting aspects of a problem, however, are our own contributions to it, the part that we play in making it a problem, because that is the part we can control best.

Had the little wife called her husband from the supermarket *before* she bought the rock hens, to ask, "And what did our motorcycling expert have for lunch today?" all would have been well.

What Are We Looking For?

Like fresh fish, information spoils with time; unlike fish, canning or freezing the information does not help it one bit. One of the great virtues of the modern computer is its ability to maintain a living, constantly changing, data bank. There are ways of interrogating a computer to ensure that it gives you either up-dated information or none at all.

No antiaircraft gunner or submarine commander would fire at the spot where the enemy had been five seconds ago, yet we confidently buy, invent, or design to meet outdated requirements. In the information search we have to seek out the most recent information possible. Taking a hint from antiaircraft gunners, we should estimate where the target will be when our projectiles get there—what the customer's requirements will be when our product reaches his hands.

The housewife's product may be a knit jacket for her sister's baby. Unless she is an instant knitter, she will allow for the child's growth while she knits.

The professional buyer's "product" may be copper for a pilot run of equipment which is to be manufactured in quantity later when copper may be scarce and aluminum abundant.

Foresight is equally important in design. Designers of topcoats for young women must anticipate trends in the march of hemlines. Foreseeing that both skirt *and coat* would approach the girl's equator in the winter of 1969-1970, they designed the maxie, an extreme full-length coat which appeared on the market at the exact time that winter's cold attacked defenseless thighs.

This example illustrates the predictive potential of information (Gabor), but freshness and predictive potential are only two of the many characteristics of the information we are looking for; selective power and truthfulness are others. Good information is true information, bad information is false information. Truthfulness measures the quality or goodness of information.

We noted earlier how self-deception can impair the quality of information. To reduce deception during the search we have to look deliberately and industriously for the data we would rather not find and ruthlessly check the happy facts we are so anxious to uncover.

The information dilemma, also mentioned later under Adequate Information, refers to the quantity of information in terms of the time available to examine and collect it. Collection includes culling useful information in order to improve the signal-to-noise ratio.

In addition to freshness, predictive potential, selective power, quality, and quantity of information, we have to consider its identity; that is, we must be able to recognize what we are looking for when we come across it. In what form will it appear? Are we looking for a few or many items? Are they constant or changing?

The expected benefits from the information we seek lie somewhere between the point of inaction defined by the phrase, "We cannot act now without more facts," and the point of last chance, "Act now or miss the boat." These are the horns of the information dilemma (Figure 5-1) that determine the content of a data-basis for action.

How to Look for It

The information phase of the value-analysis job plan does not really end until the profits are in the till, but we cannot afford to wait beyond the point that enough data for action has been collected. Therefore the data-basis is planned in answer to the question: "What information do we *have* to have before we act?" Once we have that data, we can fly with it. Now, how to get it.

Before organizing the search we must examine our resources. How much time can we spend? How many people can we use? How far can we send them? How much equipment is available? Applying the answers to the required data-basis governs the scope of the search. Once it has been determined, we can divide the area into sectors or possible locations. Obviously the search will be most efficient when we start with the most likely sector.

Weisselberg and Cowley* offer pertinent examples. The first one is about finding a needle in a haystack, "When you dropped the needle you

---

* From *The Executive Strategist* by Weisselberg and Cowley. Copyright 1969 by McGraw-Hill Book Company. Used with permission of McGraw-Hill Book Company.

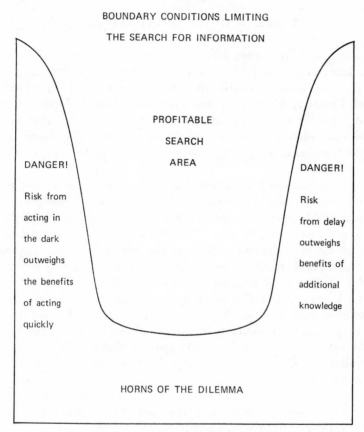

BOUNDARY CONDITIONS LIMITING

THE SEARCH FOR INFORMATION

PROFITABLE

SEARCH

AREA

DANGER!   DANGER!

Risk from   Risk

acting in   from delay

the dark   outweighs

outweighs   benefits of

the benefits   additional

of acting   knowledge

quickly

HORNS OF THE DILEMMA

Figure 5-1   The information dilemma.

must have been at one side of the stack." This statement cuts the stack vertically into a near and a far sector. "It isn't likely that you could have dropped the needle above your arm level." Now the haystack has been sliced horizontally, into upper and lower sectors. The lower near sector is the point to start the search.

The second example they offer is about a male dog who has disappeared at a time when a neighbor's female dog is enjoying a period of romance: "Where do you look first?"

Two more examples from the same fine book illustrate important search factors: (a) eliminating repetition, "By setting aside each batch of hay that you examine, you avoid looking at the same hay repeatedly," and (b) suitability of search methods, "A large magnet might pick up the lost needle in seconds."

THE PRINCIPLE OF ADEQUATE INFORMATION

This principle derives from a clear-cut pattern in nature. Motion into danger leads to destruction; motion away from danger and toward the sources of life leads to continuing life. The cycle consists of sensing, remembering, screening, and comparing thousands of inputs to select meaningful information. There is seldom enough of the right kind of information and always too much of the wrong kind. This signal-to-noise ratio yields an inadequate image that must be completed by the imagination and memory.

We do not have to see all four legs of a table to know that it is a table nor all the stars of the Great Dipper to find the North Star. We see a part and imagine the whole. Such creative visualization is the key to pattern recognition, but, alas! it is also a source of self-deception.

Perhaps the most difficult obstacle to the proper use of information is self-deception and emotional bias. The information necessary for successful value analysis must be gathered to *include the information that we do not like and to question the information that we do like*. All data must be graded for freshness. How recent is it? Is it new or is it something we already know? The competence and reliability of the source must be assessed, and the item of information must be checked for completeness and availability in a form usable by the task group. Table 5-1 summarizes the search.

We used to say, "Get all the facts!" You can't get *all* the facts. If you waited to get them you would miss the boat. "Get as much as you can" will not do either. It sets no limits.

We all live between the horns of an information dilemma, between the dangerous consequences of inadequate information, on the one hand, and the sad consequences of waiting for complete information on the other.

Coping with the Information Dilemma

One way to avoid damage from the horns of a dilemma is to locate the point of each horn with precision. One is the risk from acting in the dark which outweighs the benefits of early action. We can move safely away from this point toward the other, the losses from delay which outweigh the benefits of additional knowledge. Getting all the facts really means getting all the facts within these bounds. Setting the bounds of adequate information is a task as important as getting the information itself (Figure 5-1).

Table 5-1   Summary of Search Factors

WHAT ARE WE LOOKING FOR?

Information Characteristics

   Freshness
   Predictive potential
   Selective power
   Truthfulness
   Good signal-to-noise ratio
   Distinguishing marks

HOW TO LOOK FOR IT

The Search Situation

   Data needed for action
   Resources for the search
   Scope of the search
   Assigning search sectors
   Sequencing search sectors
   Finding suitable search methods
   Control to avoid repetition

Appraising the Quantity of Information

At the start of a value-analysis project a close look at the information available almost always reveals the need for additional information. Within the bounds just mentioned, the group must determine (a) the additional information it needs to complete its information requirement and (b) the information it can afford to pass up in order to complete the task on time. The group must then set machinery in motion for getting the additional information and for supplementing existing information.

Whenever possible, members of the task group should supplement the information available with personal visits to the engineering, production, styling, and marketing facilities of their own company and to the facilities of key vendors.

To avoid waste of priceless time the actual improvement effort should not begin during the information phase. The task group should have a clear idea of what it has to improve before it begins improving it. Getting a few answers to the right question works out better than getting many answers to the wrong questions.

Turning Information into Gold

To improve value we have to provide greater benefits for a given cost or the same benefits for less cost, not *greater* benefits for *less* cost—unless we can find a mine that yields resources at no cost.

Such a gold mine does exist, and by systematically developing the right kind of information we can home in on it. This is a dynamic gold mine in which gold flows down the hidden crevices of waste. Very real wealth can be rescued from this fate.

Eliminating friction and heat and electrical losses releases all the power these baddies were using up. Eliminating needless parts releases materials for better use, cuts down weight, and adds to shipping space. It may also increase reliability—fewer things to go wrong. It often improves performance—better use of power.

STEPS IN THE SEARCH PATTERN

Verifying Requirements and Constraints

The audacity to question specifications, the tenacity to track them down, and the skill to survive the experience are essential value-analysis traits. Most requirements and constraints originate within the inner sanctum of one particular discipline: engineering, procurement, contracting, . . . whatever. The priests of that discipline prepare specifications in the pious hope that the profane world outside will comply without unreasonable waste of time and money.

The "profane world" is made up of the other disciplines that contribute directly to product value.

Management expects, and obtains, from value analysis a nonsectarian look at all "musts" and "no-no's" that govern the value of a product. The profane world, either in a task group or one pagan at a time, can render a service to the spec writer by showing him how his specifications affect their tasks.

Many specifications, however, are "handed down from above"; some are "sacred" and others "untouchable." But "Above" is not stupid. If they can get a better product for the same money or the same product for less money, they will listen. In any case, your own management expects you to uncover pitfalls in the specs.

I have been on the "Above" side of the specs, as this vendor's phone call will reveal. "I've been trying for weeks to talk to you," he complained. "How can I get to see you?"

I sputtered, "Out of town. Tight schedule. What's the problem?"

Patiently, he explained, "You know those gookuses that won't fit inside the supergookuses?"

I knew.

"Well, you are paying through the nose, just to have that problem."

"Your plant or mine?" I interrupted to arrange the meeting.

I had not questioned the specs. I had not followed the advice of Larry Miles to challenge everything!

A frequent cause of wasted resources is the requirement for greater benefits than can possibly be used. Every specification must be examined in the light of the customer's needs or desires in order to make sure that we are not providing more than he wants in one area and less than he wants in another. Every statement that "the customer will not let us do this" or "the customer—or our home office—will not accept that" must be verified, for the absent customer or the remote home office is the last resort of the man who does not want to do something and needs a good reason for not doing it.

Validating Tolerances and Allowances

Validating tolerances and allowances calls for investigation to determine the need *and* adequacy of all safety factors, derating factors, power reserves, tolerances, allowances, and time cushions. Meant to offset actual uncertainty, allowances and reserves are often increased to offset the additional uncertainty created when the pressure of work forces designers to guess because they have no time to calculate.

Sometimes it works the other way. Tight tolerances are opened up in the interest of economy only to wipe out all manufacturing savings through rework. By *validating* I mean making sure that tolerances and allowances are neither too tight nor too loose. They should be continuously readapted to changing conditions of supply, to changing technology, and to increasing experience with a given product.

Identifying Potential Pay Dirt

A book that has taught me much about the value of information is Grayson's *Decisions Under Uncertainty*. He explains the role of preliminary information in drilling for oil. Not until you actually drill do you know whether you are going to get oil, gas, salt water, or a dry hole.

Why spend a lot of time and money on seismic studies, geological investigations, and gravity measurements when you will not really know until you drill?

Obviously because the preliminary information points to where the oil is most likely to be. You can spend all your money on information, none on drilling, and get no oil, or you can drill at random, starting in your own backyard, and maybe find oil. One of the most difficult decisions in oil exploration is how much time and money to put into information and how much into drilling.

We have the same problem in value analysis. We really do not know where the gains are until after implementation. Just as drilling for oil generates richer and surer information than surface studies, so does the analytic phase of the job plan reveal unsuspected possibilities, but neither in oil exploration nor value analysis can one count on unlimited resources, and so we must hit the likeliest areas first. Mind you, *first*—not exclusively—because the analytic phase provides a rewarding opportunity for random exploring. To hunt lions you have to go to a country where there are lions.

Rules vary for different animals and for different classes of product. Looking at a commercial TV chassis, a value analyst, on loan from another plant, announced, "We are going to work on the brackets, cable clamps, and hardware."

"Why?"

"Because the electronic components and the circuitry do not lend themselves to value analysis" (!).

The truth is that in commercial TV the electronic components and the circuitry yield five times as much as the mechanical parts of the equipment. I say yield because the primary value targets in a household TV set are not cost savings but improved safety and reliability, with cost following as a close third. The correlation, however, is the same. The newer components work better, are safer and more reliable, *and*, with up-dated circuitry, fewer components provide a better function; so all together the improved product costs less—five times less than when effort is applied to the mechanical elements alone.

Updating and Looking Ahead

The faster our technology grows, the greater the opportunities for updating our thinking on new customer requirements, new methods, and new materials. But what is new today is not good enough. Our competition may be ready for tomorrow, and we have to design around tomorrow's needs, tomorrow's methods, and tomorrow's materials.

Moreover, the evolution of a product is such that minor defects—each of which is declared unimportant—together can throw the product out of the market. Continuously updating a good product, on the other hand, can keep that product in the lead. Systematically anticipating trends can ensure product leadership tomorrow.

Again and again we must ask, are we taking full advantage of

the state of the art
new materials
new techniques
changes in customer needs
new markets

Does the present approach take advantage of

new manufacturing methods
new capital equipment

Processing Information into Knowledge

At this point the mixture of relevant skills begins to prove its value. All the verified, validated, and updated facts that have been gathered are jointly appraised by the whole group, not as they affect each department but as they affect the business. This basic knowledge is summarized, but the door is left open for ferreting out the additional information that will undoubtedly be required as the team enters each new phase of the job plan. Before going into the analytic phase, for example, the most promising areas of each project should be selected. Such selection calls for an appraisal of the information already available and for determining what else the team must know in order to make acceptable recommendations. The selected portions can then serve as a basis for comparison with the many other opportunities that the analytic phase invariably reveals.

Selectivity in the Request for Data

There *is* such a thing as too much information. Information that is not relevant is simply noise. Slightly useful information may well obscure the really significant facts. Request only the information that appears to be really significant for the value study, but do follow hunches and do ask for information that appears to be significant to you even though others may disagree.

Conduct an organized search for problem areas and improvement opportunities. Write down any ideas that turn up during the search but do not expose them to criticism until the creative phase.

Table 5-2   Check List of Essential Data

Identify the Project

1. Is it a product ( ) or a service ( )?
2. Product line and type number _____

3. Service to be analyzed _____

Define Scope of the Study

1. What is it expected to do? Should we analyze the entire project? Yes ( ) No ( )
2. If not, what part of it? _____

3. What techniques may we consider? Only those that utilize our existing capital equipment and present skills ( ) or any type of equipment, skills or process ( )
4. What are the most significant features of this project? _____

5. What is the state of development? _____

Determine Quantities and Product Life

1. Where is this item used? _____

2. How many go into the next higher assembly? _____
3. Where else is this item used? _____

4. How many units will be produced over the next two years? _____

Learn Marketing Requirements

1. Is this what the customer *really* wants today? _____

2. What are the customer's desires; what will he want tomorrow? _____

Table 5-2 (*Continued*)   Check List of Essential Data

3. What will contribute most to sales? (rank the items with the help of Marketing and Styling):

greater safety	( )
greater reliability	( )
improved performance	( )
added capability	( )
improved appearance	( )
easier maintenance	( )
lower overall cost	( )
lower sales price	( )

Review the Cost Data

1. Fill in the universal cost chart in Figure 5–2
2. Is the cost information in shape for use during the workshop?
   If YES ( ) go on to the section *selectivity in the request for data.*
   If NO ( ) use a copy of this table to request additional data on
   Customer requirements _____
   Design engineering _____
   Documentation _____
   Tooling _____
   Raw materials _____
   Purchased parts and components _____
   Manufacturing costs _____
   Engineering follow _____
   Service under warranty _____
   Customer's cost of _____

   operation _____
   maintenance _____
   repair _____
   downtime _____
   depreciation _____

Explore the sources of random information that appear during the planned search but return to the plan. When the planned search is completed, do some random exploring for unexpected morsels of information.

PINPOINT THE SIGNIFICANT FACTS

A summary is an abbreviated review of all the data. What the team needs at this point is not a summary but a distillation of useful data.

This copy of the Cost Chart shows (indicate which): actual cost ( ), estimated cost ( ) of the present approach ( ), of a proposed approach (identify):_____

Name or Description	Part or Drawing Number	Quantity Over 12 Months

NONRECURRING COSTS PER UNIT		RECURRING COSTS PER UNIT	
Tools	$	Raw materials	$
Facilities	$	Purchased parts	$
*MHX (%)	$_____	Subcontracts	$
TOTAL MATERIALS . . . . . $		Other materials	$_____
Engineering	$	Subtotal	$
		MHX (%)	$_____
Drafting	$	TOTAL MATERIALS . . . $	
Model Shop	$		
Testing	$	Direct labor Hours	$
Qualification	$	Average rate	$
Setup	$	Direct labor	$
Documentation	$_____	cost	
		ESE (%)	$_____
Subtotal	$	TOTAL LABOR . . . . . $	
†ESE (%)	$_____		
TOTAL LABOR . . . . . .$		OTHER . . . . $_____	
OTHER . . . . $_____			
Total nonrecurring		Total Recurring Costs	$
costs . . . . $			.$
		TOTAL UNIT COST	$

*MHX = material handling expense, †ESE = employee service expense; no other overhead.

Note. Copies of a chart similar to this one, but reflecting your own company's costing methods, should be provided in abundance to all task group members.

Figure 5-2 Universal cost chart.

A good fractioning still, used in Scotland and Canada, selectively distills potable spirits, leaving out some "character" and many hangovers. Old-fashioned pot stills, used in the Ozarks, the New Jersey Pine Barrens, and Ireland, leave all the character and some hangovers in the whiskey.

The "character" that can make team members heroes at the expense of other people's mistakes must be distilled out, together with the hangovers arising out of exposés. The limited information selected must become, by the very act of selection, much more powerful in its effect than the unlimited information rejected, which brings us to a more thorough analysis of . . .

THE PRINCIPLE OF SELECTIVITY

There is a subtle difference in meaning between the words *selection* and *selectivity*. *Selection* means choosing one or more from a larger group of options. Akin to *choice*, selection is the final step in the decision process. It is also an inherent part of the search for information. *Selectivity*, on the other hand, refers to a systematic *policy* of selection or a systematic pattern for selection. Solomon applied the principle of selectivity in fortifying the hills of the Negev, in the location of his copper smelters, and undoubtedly in family life with his 700 wives and 300 concubines.

Selective Nature of Information

You are at a neighborhood cocktail party. Everybody is talking at once, yet you catch the sound of your name. *That* is information; the rest is noise. But the information is coming from a room where there is no food, so you continue moving toward the smell of little sausages wrapped in crisp bacon. There are innumerable other aromas floating around—the tang of Italian vermouth, My Sin, rich pipe tobacco, Arpege, Angostura bitters. To a hungry man at 6:00 P.M. the smell of little sausages wrapped in bacon is information; the rest is noise.

Sniffing hungrily, your nose selected from among a set of promising aromas the one most closely related to your needs at the time.

The principle of selectivity applies to the gathering, transmission, storage, and retrieval of information. In all these tasks, except transmission, the pattern for selection is governed by the usefulness of information. Transmission is another story. To be worth transmitting the information must already have been selected for usefulness.

To be useful information must be true, relevant, and fresh, offering the opportunity to predict and select.

Weisselberg and Cowley's needle in the haystack is a good example of this kind of selection. The vertical partitioning selected the near side of the haystack and the horizontal partitioning selected the lower portion of the near side; then the magnet selected the needle.

Selectivity in the Choice of Tasks

The products to be analyzed are usually selected under the guidance of the general manager and the responsible product or project managers. These executives, together with the value specialist, his boss, and the plant training manager, constitute a planning group for the value-analysis effort. Their selection of projects (see Figure 5-3) is usually based on

potential gains
technical feasibility
availability of information
availability of skills
timeliness
probability of implementation

Selectivity in the Choice of Skills

Once the planning group has selected *WHAT* to analyze it decides *WHO* will do the analysis. This selection must achieve the best combination of skills to fit each project.

Figure 4-1, in the preceding chapter, shows one way of filling the slots in each task group. This chart can be used to organize a single task group or to plan a full-fledged workshop with a number of teams.

Selectivity in Conducting the Analysis

A purchasing manager often hands a bill of materials to a buyer, with the words, "Eighty per cent of the cost is clustered around 20% of the items. Work on that 20%." This type of inbalance was first noted (1896) by Vilfredo Pareto, the Parisian born, Italian engineer who taught economics at Lausanne. An experienced and competent mathematician, Pareto plotted the distribution of income in many lands. He discovered that despite widely differing social and economic conditions a few of the people raked in most of the money.

The disproportion—though not always applicable to income—became known as "Pareto's law of mal-distribution," and it keeps popping up

SELECTION CRITERIA / CHOICES	EXAMPLE			
POTENTIAL What is the proposed improvement worth? Gross expected gains LESS cost to implement	$1200 200			
	1000			
FEASIBILITY What are the chance it will work as planned? 100% = sure to work	x 90%			
	= 900			
INFORMATION Have we or can we get all the information we need? 100% = complete information	x 80%			
	= 720			
SKILLS How suitable are our skills for tackling this project? 100% = fully suitable	x 100%			
	= 720			
TIME Have we enough time? 100% = ample time	x 90%			
	= 648			
CHANCES OF IMPLEMENTATION 100% = certainty	x 80%			
NET EXPECTED GAINS	= $518			

Figure 5-3 Project selection and concentration of effort.

everywhere, especially in the potential for value improvement. One small subassembly of the product or one phase of the service may have much greater improvement potential than all the rest put together. The principle of selectivity calls on the value task group to search for such inbalance and to concentrate its efforts on the area in which its particular skills promise greatest yield for the time, information, and tools available.

VI

ANALYTIC PHASE

Value analysis is a sharp tool. Using it on the wrong material can blunt the tool and discredit the user. A costly team of specialists assigned to work on a low-cost, low-volume product which is nearing the end of its run can hardly produce results commensurate with the price of their skills. Assignment of such a low-yield goal reduces the effectiveness of value analysis in the eyes of management. The tool loses its edge.

In other situations the yield may be there but it may be missed altogether. A mountain of talent can labor to bring forth a mouse, which can happen when a blue-ribbon team of biochemists, physicians, and medical statisticians decides to work on the physical manufacture and packaging of their company's drugs. Conversely, a production broodmare can labor and bring forth an elephant who suffers from birth defects when a factory team decides to work on the formulation of the company's drugs.

At the start of the analytic phase the projects have been selected and promising areas identified. It is up to the teams to appraise their own skills in order to set goals that match those skills. An analyst, working alone, also has to appraise the skill centers available to him for consultation and set goals that utilize those skills. Setting goals is governed by . . .

. . .

THE PRINCIPLE OF DIRECTION

And all I ask is a tall ship
and a star to steer her by.

John Masefield

137

One definition of direction is "guidance or supervision of action or conduct." This guidance, however, is dependent on direction in the broad sense of setting a course. Before setting a course, the line executive has to decide where he wants to go. He can then guide his subordinates in the right direction. The value-analysis task group has a more subtle task. Before it can persuade equals and convince superiors to follow a given path, it must show them that the proposed path also leads to *their* immediate goals.

For an excellent illustration of goals and objectives, I have to take the reader to an old hacienda in the Colombian Andes. As you leave the hacienda to go on a trip, you have to pass under an archway decorated with three figures: a stately pilgrim carrying a staff, a young man with a large fish over his shoulder, and a dog running merrily beside them.

The group illustrates the Book of Tobit, one of the most delightful episodes in the Apocrypha. Represented in the sculpture are Tobias and his dog and the fish he caught and the Archangel Raphael, the first consultant on goals and objectives.

GOALS AND OBJECTIVES: A CASE STUDY

PARTICIPANTS

TOBIT, a patriarch	A MAN-EATING FISH
TOBIAS, his son	SARA, an enchanted maiden
RAPHAEL, an archangel	ASMODEUS, a demon

LOCALE

Nineveh, the River Tigris, Ecbatana in Media

Tobit, one of the Jewish captives in Nineveh, has been deprived of his livelihood, loses all his property, and goes blind. All he has left to give his son Tobias are 10 talents of silver he had left on deposit with a kinsman in the distant town of Rages in Media, but the boy, Tobias, does not know how to get to Rages and Tobit is too old to remember the way. Discouraged, Tobit prays to be gathered to his fathers, for he has lived too long.

In Ecbatana, capital of Media, the fair and good maiden Sara also prays for death. She has been enchanted by the demon Asmodeus, who kills her bridegrooms on their wedding night.

In heaven the archangel Raphael is assigned the task of relieving the troubles of Tobit and Sara. First he is given the general objective, heal Tobit and Sara, then the three goals leading to that objective: the explicit goal of giving Sara as a wife to Tobias; the implicit goal of binding the demon Asmodeus because a wedding is no good if the groom is killed; and the explicit goal of healing the blindness of Tobit.

Raphael appears in Nineveh in the guise of an experienced traveler. He is brought to Tobit who, forgetting *his* goal, begins a lengthy interrogation, but Raphael stops him. "Dost thou seek a tribe, or a family, or a hired man to go with thy son?"

Tobit then hires Raphael as a guide for Tobias and blesses them both. So they set out and the young man's dog goes with them.

When Tobias approaches the banks of the Tigris, a great fish leaps out to devour him. Man-eating fish who threaten to devour one's clients are, as a rule, a problem, but to Raphael the fish is a target of opportunity. "Take the fish!" he orders, and as the young man draws the fish up on the land the archangel adds, "Open the fish and take the heart and the liver and the gall, and put them up safely."

They eat the rest of the fish and continue on their journey. Near the city of Ecbatana, in Media, Raphael suggests that they stop at Sara's home. Her father is Tobit's kinsman and, according to the archangel, Sara should be Tobias' wife by right of inheritance.

The young couple fall in love at once, and Raphael attains his first goal by arranging the marriage. In the early evening the wedding feast turns to gloom as the spirit of Asmodeus darkens the sunset.

Summoning all his courage, Tobias enters the wedding chamber, where he is sure to be slain by the demon; but, following the instructions of Raphael, he drops the fish heart and liver on the embers of the incense brazier. At once a cloud of black smoke spreads toward Asmodeus and drives him to the uttermost parts of Egypt. There the archangel binds him securely, thus attaining his second goal.

While the young people enjoy their honeymoon, Raphael takes Tobit's letter-of-credit to the town of Rages and returns with the 10 talents of silver. Then the archangel and Tobias, with Tobias' bride, return to Nineveh, the young man's dog going with them.

Arriving in Nineveh, the young man puts some of the gall of the fish on his father's eyes, whereupon the sight of Tobit is restored. Having attained his third goal and realized his general objective, to heal Tobit and Sara, the archangel Raphael departs, leaving the young man's dog to look after his master.

Summary

Raphael is the healing archangel. Healing is his general objective— what he tries to do all the time. Companies, too, have general objectives. What a company tries to do all the time is to survive. In order to survive it needs revenue; then come other objectives, such as profit, growth, product leadership, and a share of the market. Leading toward these objectives, operations within the company have their own goals.

The explicit goals of value analysis are product improvement and cost reduction done in relation to each other. Implicit goals are better information, more accurate communication, and smoother interaction among specialists; and then there are targets of opportunity, those unexpected goals that are suddenly revealed and turn out to be sheer gravy.

It also turns out, however, that the seldom-mentioned implicit goals of value analysis deserve the most attention. Why? Because the explicit goals—product improvement and cost reduction—are also the goals of other departments. Value analysis simply relates these goals to one another, changing divergent into joint effort.

Importance of Information and Communications

Getting the product improvement and cost reduction experts to work with instead of against each other cannot be accomplished by preaching, training, or threats. It is a matter of providing a practical means for the early exchange of information. Then the goals can be changed from "the most of the best of everything," and "the least possible cost" to the best relationship of worth to cost for a given purpose—value!

The principle of direction applies to Raphael's plan in the definition and sequencing of his tasks. He knew his destination, understood his goals, and defined his objectives before he set out to solve problems and exploit opportunities. He knew where he was going!

Before we begin building bridges or digging tunnels we should know where we are going and why. The *why* is determined by the value we set on the objective and the *where* puts us within reach of attaining that objective. Both can change at any time, and we must be ready to change direction accordingly. Applying the principle of direction relates the value program to the objectives of the company, establishes specific goals for the task groups, and shows the participants how to reconcile personal, departmental, and company objectives.

TASK GROUP GOALS

Improving the value of a company's product certainly contributes to company objectives, but what is seldom realized is that participation in the task group improves the value of the employee himself. He learns to look at a product through the eyes of other specialists and he establishes useful personal contacts in their departments. Realizing that the task group is only an integrating force, he learns to respect the people responsible for the product and to consider the cost of resources that go into the product.

SHARP TOOLS ARE DANGEROUS

Misused, value analysis can do serious damage. A good idea, no matter how difficult its development, seems obvious to everybody, once it proves successful. In business and industry someone is invariably pilloried for not having "thought about it in the first place."

Who is the victim? Usually the person who did all the groundwork which made it possible for the idea to come into being. This victim of sanctimonious hindsight is often the man responsible for the product. Hurt him and value analysis becomes as welcome as the Black Death.

THE PRINCIPLE OF RESPONSIBILITY

The word *authority* comes from the Latin *auctor,* which means *doer*; hence the close connection with the word responsibility which comes from *responsum,* meaning *answer.* How can a man answer for his acts if someone is "helping" him against his will? This problem must be faced sooner or later by every supporting function.

The smoothest, fastest, and simplest way to operate a value task group is to identify the man responsible at each key decision point in the life of the product and to go to work for that man. Most decisions involve risk. Risk calls for judgment and courage. We cannot afford to be lions of courage with someone else's guts. For this reason the gains in product worth and savings in product cost should always be credited to the risk taker.

In general, the decision maker can be identified by reviewing in your mind's eye all the people who influence product value at any particular moment. If you are reasonably perceptive, you will see a hemp rope around the neck of one of them. He is the one who hangs if things go wrong. He should be your chief source of information and should have the final say-so on your recommendations.

Putting the Tools to Work

Now that we know how to care for our tools and how to avoid their misuse, we can plunge into the analysis itself.

ANALYZING THE FUNCTION

The simple question, "What does it do?" takes us from the structure to the function of a product. It deals with matters of fact and definition. "What *should* it do?" deals with matters of value and judgment. "Why?" brings into play goals and objectives. "How?" initiates the search for a path to their attainment. It calls for direction and method. The road signs and the milestones now say VALUE instead of LEAST COST, and the method is analysis.

What Does Analysis Mean?

To analyze is to separate the components of a whole for examination or study. Breaking down a problem into its parts is analysis; examining the individual components of a product is analysis; breaking down total performance into component functions and subfunctions is analysis. The word comes to us from the Indo-European root *leu, leus,* to loosen or untie, through the Greek *-lysis* from *lyein* to loosen. *Ana-* is the Greek equivalent for our *un-,* in *untie, untangle, unravel.*

What we have to do in the analytic phase is to untangle the confusion between function and structure, unravel the skein of functions, and loosen the conceptual bounds that hold the project together; then we can take everything apart and examine it.

Our target now is the function or functions of the product. The function is governed by the . . .

PRINCIPLE OF USEFULNESS

In the sense of *performance toward a given end,* the word *function* is tied to the concepts of direction and end-use, and we should bear in mind that the given end is entirely within the province of the customer. He is interested in what the product will do for him according to his own standards. Satisfying a customer is the end purpose of all industrial products in the free world. He may be satisfied in terms of end-use, esteem, or both.

If the customer wants a little beauty in his life, is willing to pay for it, and wants it in living color, *that* is what we must give him! Conversely, if the God of War wants more military equipment for his money, even if the equipment looks horrible, we simply have to comply. A word of warning, however—the God of War is less austere than he pretends to be. If he secretly wants the equipment to look good and it doesn't, he may turn it down for "quality."

The principle of usefulness calls for two of the tasks already mentioned under *The Nature of Value*:

1. Finding out what the customer really wants our product to do for him.

2. Designing, making, or buying a product that will meet his needs and desires *in proportion to the relative importance he gives each one of them.*

Whoever the customer is, the product must be worth more to him than he pays for it; otherwise he would have no reason to make the exchange. In addition to giving him his money's worth, the supplier therefore has to compensate the customer for the risk and effort of buying. A good reputation, opportune delivery, and good service usually provide such compensation over and above the sale price. The sale price then remains as the minimum measure, or lower limit, of what the function itself is worth to the customer. We have to give him that much utility at the very least.

Breaking even on utility and compensating the customer for the risk and effort of buying is not enough. To be competitive we have to improve the function, reduce the sale price, or improve the way the function is performed.

DEFINE THE FUNCTION

If we take a complex idea and condense it into two words, say a verb and noun, we lose information. If we do it on purpose, we have to decide *what* information we can afford to lose. Then we are left with the most significant aspect of the product's purpose—the primary function.

It is important to bear in mind that we are not attempting a definition for the benefit of all mankind but are simply looking for one or two words that will pinpoint the primary function of the product for the benefit of the task group (Figure 6-1).

This simple definition, however, should be at the highest level of abstraction compatible with the company's business.

To raise the level of abstraction, we ask "why?"
To lower the level of abstraction, we ask "how?"

Remember, however, that we are considering the highest level of

SYSTEM, ASSEMBLY, PART, OR PROCESS	WHAT DOES IT DO?		DEGREE				ESTIMATED COST
	Verb	Noun	1st	2nd	3rd	4th	

A definition has different purposes in varying situations. The purpose here is to pinpoint the most important aspect of the function by boiling down the concept into a verb and noun. The primary function is checked off in the first column under DEGREE, less important functions in the other columns. The COST column will serve later to determine if cost is proportional to degree of importance.

Figure 6-1 Define the function.

abstraction compatible with the company's business. This constraint is necessary because overabstraction can take you clear out of the ball park. Here is an example. A blue-ribbon value task group decided to have a little fun during the preliminary analysis of a future air defense radar.

At the presentation to management, when each team offers its recommendations before the general manager and his staff, the team spokesman turned to the chart which defined the function as DETECT TARGETS.

"So we asked why?" he explained, showing the next chart: DESTROY TARGETS.

He went on, "Why?"

ENEMY TARGETS.

"Why?"

OUR NATION IS AT WAR!

"Why?"

THAT IS ONE WAY NATIONS RESOLVE CONFLICT!

"Is that the best way?"

NO.

"What else will do?"

Fortunately, the first two teams had come up with good recommendations; still, some of the managers were looking puzzled and the general manager's lips were quivering like the lips of the volcano Krakatau before the explosion.

"Gentlemen," the spokesman continued, "Here is our recommendation."

The next chart showed something that resembled a radar dish with a thin pedestal, but it was full of birds!

"An olive tree, gentlemen," the spokesman continued, "because the olive branch is an age-old symbol of conciliation, and the doves of peace—"

The explosion occurred. "I'm not running a damn olive grove full of pigeons," roared the boss. "I have a plant and machinery to make *radars* and I *thought* I had engineers to design them." He stopped vibrating and added, in a dangerously level tone, "You guys show me a good radar design or your"—he used an anatomical term—"will land in the middle of the street *right now!*"

"Very well, sir," the spokesman replied, flipping more charts to reveal an excellent radar design.

You can risk overabstraction when you are prepared with an alternative really likely to suit the company's resources.

Keeping the Definition Simple

One can define a function in two words and get terribly complicated. Chemical terms are an example. What we need are simple down-to-earth definitions readily understood by the task group.

When Dave Johnson* headed the value program at Whirlpool Laundry Products, he assigned a task group to look at all the peripheral aspects of their washing machines. One of the items considered was the disposal hose. This hose is possibly the ugliest product of western civilization, a piece of tubing, part flexible and part stiff, which hangs like a discarded gut on the edge of the laundry tubs, the other end being connected to the washing machine.

Whenever the man of the house approaches the laundry tubs for any purpose, this monstrous device disgorges enormous quantities of evil-looking tepid water.

It is only a bent hose which looks like a king snake beginning to decompose, but it was costing the company 23¢ to put the bend in the hose. A steel mandrel had to be inserted by hand; the enforced bend, mandrel and all, had to be baked in an oven and then the mandrel had to be taken out.

The assignment was to analyze the bend in the hose. It was costing too much for just a bend.

"What does it do?" asked Dave Johnson.

A tired, but cultured voice, from the rear of the room rolled out the words, "It alters the direction of water by a hundred and eighty degrees."

"No, no," someone protested, "more than water goes through that tube: detergent and dirt in suspension—what is the name of that fluid?"

A summer student, who had brought in a package of drawings, gave the fluid an unscientific but apt name.

Dave Johnson interjected, "Watch your language, son," and then,

"Oh, come on, fellows," he pleaded, "what does this thing do? Give it to me in a verb and a noun."

Silence.

Then the summer student spoke up. "Heck, I can tell you what it does. It bends water."

"Bends water?" answered Dave. "Fine, we'll fly with that. Everybody understand it? . . . O.K."

The workshop moved on. Once the task group understands the function, nothing is gained by lengthy philosophical discussions on the precise verb and noun to describe it.

---

* Currently Corporate Director of Value Engineering, Magnavox Company.

Typically, the function is defined in an active verb and a measurable noun. Examples are

CONDUCT CURRENT
TRANSMIT TORQUE
CUT GRASS
AGITATE CLOTHES
SELECT CYCLE

From What Is to What Should Be

During the 1940's and 50's the concept of value entered value analysis through the question, "What *should* it cost?" This was natural enough in a discipline that developed from solving problems of scarce resources. The word *should*, however, is the magic word that evokes the Spirit of Value. He is a spirit as powerful as the civilizing heroes of ancient days; yet he is as modern as tomorrow. Pointing to a washboard, he asks, "What should it do?"

Behold! A washing machine!

The question, "What does it do?" identified the functions of washboards and electric fans in their day.

"What should it do?" directs the thinking of buyers, marketing men, designers, and inventors toward the next step—washing machines and air conditioners. Ask the question again and we think of washer-dryers and air purifiers.

The business goals set for a particular value-improvement effort determine how much weight should be given to each of these two questions.

If we are making washboards and they are selling well, the questions

"What does it do?"
"What does it cost?"
"What should it cost?"

might be all that is needed.

On the other hand, if people are no longer buying washboards, or if another nation can produce washboards and sell them in our market at less than our manufacturing cost, it is time to ask, "What should it do?" and move ahead.

The functions about which we ask these questions are usually ranked as follows:

Primary, Secondary, and Lower Degree Functions

This classification has become virtually standard nomenclature in value analysis.

The Primary Function. Essential to the business success or military effectiveness of a product, the primary function is the value-analysis designation for the main reason why a product is made or a future product planned. This function is necessary to make the product work or sell. In a lawn mower the primary function is *cut grass.* Yes, grass *is* a measurable noun. It can be measured in square feet of lawn, pounds of grass cuttings, or length of grass blades the mower can cut.

The customer may not think in terms of square feet, but he knows that a 30-inch mower gets the job done in less time than an 18-incher.

It is important to bear in mind that the primary function is the reason why the customer buys the product.

Primary or Secondary? Cut grass is the primary function of both hand and power mowers, but if your company makes only power mowers then *provide power* is another primary function without which the mower will neither work nor sell—or is it? Need this situation put us in a quandary?

There is a story about a hired man who was sent to the root cellar of a farm to separate the good potatoes from the bad ones. When the farmer went to the cellar at the end of the day, he found three piles of potatoes. There was a small pile of outstandingly good potatoes, a small pile of rotten potatoes and a large pile of . . .

"Neither one nor the other," explained the hired man, "and the decisions damn near killed me."

We don't have to make life tough for ourselves by discriminating, with great precision, between primary, secondary, and lower degree functions. Mutually dependent functions are particularly hard to label and so are equally important functions, but the purpose of the label is to determine which function to work on. If two or more functions deserve equal attention, we work on both of them, that is all.

Secondary Functions. A secondary function helps the product work better and, hopefully, sell better. Let us consider those consumers who buy power mowers. Some of them want a power mower that they can push and thus enjoy an invigorating walk; others—the golf-cart type—prefer a riding mower. Both, however, want to cut grass under power. *Cut grass* and *provide power* are then primary functions. *Carry rider* is a secondary function.

Lower Degree Functions. Features of less importance, such as *collect cuttings* in a mower that gathers up the grass it cuts, are called *lower degree functions.* What this amounts to, really, is a ranking of the functions, as provided for in Figure 6-2.

Once we have identified the function, we proceed with the analysis by identifying the benefits it provides. Figures 6-2 and 6-3 present a frame-

IDENTIFY THE BENEFITS

Examples of benefits are performance, quality, reliability, ease of operation, repair and maintenance, length of life, appearance, and early delivery.

List only those benefits likely to be affected by the proposed effort. Write them down at random.

You may want to consider only one or two significant benefits.

DETERMINE THEIR RELATIVE IMPORTANCE

Rank the benefits in order of importance

Give each benefit a verbal rating such as "most important of all, very important, fairly important, unimportant."

When feasible, and if you think it will help, assign a fractional weighting factor to each benefit so that the sum of the weights equal 1.0

Example:
0.5 + 0.3 + 0.2
= 1.0

BENEFITS	RANK	VERBAL RATING	WEIGHT
			0.
			0.
			0.
			0.
			0.
			0.
			0.

ITEM BEING CONSIDERED

Sum of the weighting factors should add up to 1.0

Figure 6-2 Identify the benefits.

149

Figure 6-3 Continuing sheet.

work for this purpose. Ranking the benefits in order of importance may be enough, but in preliminary design it is always wise to give them measures of relative importance in such a way that the sum of the measures adds up to 1. The function *transport people* by automobile can be analyzed as follows:

Economy Car	Rank	Weight	Sports Car	Rank	Weight
Initial economy	1	0.3	Speed	1	0.3
Operating economy	2	0.3	Appearance	2	0.2
Dependability	3	0.2	Initial economy	3	0.15
Comfort	4	0.1	Dependability	4	0.15
Speed	5	0.05	Operating economy	5	0.1
Appearance	6	0.05	Comfort	6	0.1

Of course, there are sports cars that are also economy cars. Instead of arguing with the ranks and weights in the example, the reader should simply change them to suit his preference; that is, if he is thinking of buying a car, for then he is His Majesty the Customer.

In Summary

Identify the function or functions of the product. Concentrate on what each function accomplishes and why. Define each function in an active verb and a measurable noun. Rank the functions by degree of importance as primary, secondary, and lower degree. Remember that words were made for man and not man for words, so don't get hung-up on the words. Note the informality revealed in the lower half of Figure 6-5.

Now that we have gone through the mechanics of analyzing the function we should consider the principles that govern the analytic phase of the job plan.

THE PRINCIPLE OF LIMITED RESOURCES

The resources available to carry out a given task or to perform a given function are limited by space, weight, dollar cost, and conflicting demands from other tasks or functions in a system.

The limitation of resources would not be a problem in a static society whose needs did not increase with time, but an exploding population whose members are continuously developing increasing needs, living in a world that is not getting any bigger, must continuously discover or create new resources. Eve, as we saw in Chapter 3, launched the food-producing revolution. And what happened? Cain and Abel and the population explosion! People multiply in just the right numbers to eat the extra food, and then some more, and we are back to limited resources.

When resources are limited and needs increase, the answer is to improve the yield from existing resources and to use each resource where it will do the most good. This concept is embodied in a principle that we must now consider in detail.

THE PRINCIPLE OF ECONOMY

Mere parsimony is not economy. . . . Expense, and great expense, may be a part of true economy. . . . Economy is a distributive virtue, and consists not in saving but in selection.

Parsimony requires no providence, no sagacity, no powers of combination, no comparison, no judgment.

Edmund Burke

The corollary to this quotation is that true economy calls for selectivity, providence, sagacity, powers of combination, comparison, and judgment. As prime minister, Edmund Burke introduced and steered through Parliament the Economic Reform Bill, passed in 1782, one of the rare instances in which a government initiated legislation to reduce its own spending power.

Burke himself made it clear that economy does not mean saving resources just to sit on them. His "distributive virtue" must not only save those resources doing the least work but it must also identify high-return opportunities in which at least some of the savings can improve the service rendered or the product manufactured.

Surprisingly, the outcome of this dual effort results in *more* resources saved.

Why?

Because a plan or design, improved with economy in mind, is often cleaner, simpler, and less cluttered.

Economy in Language

An editor hands three chapters of manuscript to one of his favorite authors. "Excellent," he says. "All you have to do now is tighten it up."

A good author then proceeds to eliminate needless detail, trim the irrelevant and the obvious, and give force to his copy by economy of language. I wish I could do the same and I do try. Hopefully the reader can do better with his own prose.

Humor, analogy, and metaphor are great instruments of economy. If you review some of the jokes that you consider really good, you will note an element of surprise. In writing or public speaking surprise, by itself, is not humor. It must be *anticipated* surprise. As the story develops, the listener is only one step behind the speaker, building up the scenario in his own mind, until the punch line suddenly reveals or upsets the whole picture—he has caught up! Now it is *his* story, too, because his mind has done more than half the work. With consummate economy, a good storyteller uses the listener's imagination, experience, knowledge, and memory as props to create a colorful production in the listeners own stage.

Arthur Koestler describes the principle of economy in language as, "Learning to create magic by the most frugal means."

Economy from the Division of Labor

An amazingly adaptable physician and successful industrialist was William Petty, who first noted that "economies from specialization and the division of labor" bring lower production costs.

Charles Babbage, one of the pioneers in the study of industrial management, wrote on this subject in 1832—quoted by Marshall (1966 ed., page 220):

The master manufacturer by dividing the work to be executed into different processes, each requiring different degrees of skill or force, can purchase exactly that precise quantity of both which is necessary for each process; whereas if the whole work were executed by one workman, that person must possess skill to perform the most difficult and sufficient strength to execute the most laborious of the operations into which the work is divided.

Economy of machinery and manufactures makes a good subtitle for this section. It is also the title of Charles Babbage's book on industrial management. Babbage anticipated most of the modern methods in plant management. It was Marshall, however, who gave us the term *economies from the scale of production*, frequently used today (see Ohlin, for example) and which stems from this often quoted statement:

The chief advantages of production on a large scale are economy of skill, economy of machinery, and economy of materials.

Marshall then enumerates the economies from specialized machinery, from improved purchasing, improved marketing, and from the acquisition of specialized skills, all resulting from large-scale production (Marshall, 1966 ed., pages 232 to 236).

The economy of high wages is the name given today to the principle first enunciated by Adam Smith and quoted by Whittaker.

A plentiful subsistence increases the bodily strength of the labourer, and the comfortable hope of bettering his condition, and of ending his days perhaps in ease and plenty animates him to exert that strength to the utmost.

Marshall agreed: ". . . generally, though not always, increased remuneration stimulates to increased exertions."

Seven years before Marshall's death Henry Ford felt that the wages he paid were not in line with his profits. Ford's wage revolution is mentioned by Professor William M. Fox in the article *Scientific Management: Taylorism* (Heyel):

Net income of $27 million was realized in 1913. Thus, in 1914, he (Ford) made an announcement which shook Detroit, and for that matter, the world: the five-dollar day.

Economy in Physical Distribution

Levy and Sampson distinguish between physical and economic distribution by describing physical distribution as an aspect of production that increases the value of goods by moving them through space and time, by carrying them from where they are not needed to where they *are* needed, by storing them when they are surplus, and releasing them as required.

Insofar as production is meant to increase the usefulness of goods, physical distribution is certainly a productive activity.

Let's say that you and a small turtle are in a life raft in the tropical Pacific. You are trying to produce drinking water with one of the solar evaporators that accompany such rafts. Three Mariner Scouts—a species of grown-up Girl Scouts who know how to find islands and fresh water on islands—approach you in another life raft.

"We found an island," they shout. "We brought you some water before you dry up altogether!"

Are you likely to answer, "Go away. I'll wait for my production water. I don't want transported water and transported company. I have a turtle, produced by my own fishing industry."

Chances are you are not that obsessed with the semantics of production. Realizing that physical distribution is the very backbone of trade, you accept the water, offer them some of your ship's biscuit, and introduce them to the turtle.

Economy in Organization

The great mining and metallurgical consortium of Commentry-Four-chambault Decazeville (Comambault) was on the verge of bankruptcy when Henri Fayol took over as chief executive. Not only did he turn the giant company around, but his excellent staff and their planning and organization became a subject of serious concern to the German Imperial General Staff and to the intelligence branch of the Krupp works in Essen; it was simply not right for Frenchmen to be that systematic, that thorough, that far-sighted, and that successful.

The Germans were right, too, because Comambault contributed substantially to allied victory in the 1914–1918 war.

When the war had been won and Comambault's financial position was impregnable (Heyel), Henri Fayol retired at the age of 77, and in answer to a veritable clamor for an explanation of his methods he republished a book—originally written in the darkest years of the war—that outlined the master plan which had turned Comambault around. This book divided the management task into Forecasting, Organization, Command, Coordination, and Control, all described under the title of *Administration Industrielle et Générale.*

Economy of Thought

"Analysis," argued Professor A. G. Kaestner of Leipzig and Göttingen, "is a superior heuristic method because of its power and *economy of thought*" (Boyer). The italics are mine. We are not interested now in the sterile argument over the merits of synthesis versus analysis in geometry, although we might note in passing that one of the arguments against analysis was that it is too easy!

The analysis that Professor Kaestner was defending was the algebraic approach to geometry of René Descartes, an approach that *does* provide economy of thought.

Ernst Mach, the physicist who demonstrated the principle of equivalence between inertial and gravitational forces, maintained that the purpose of theory in science is economy of thought, in what he called *Prinzip der Denk-Ökonomie.*

Economy in Nature

We are all familiar with the law of conservation of energy, a part of natures' system of cost accounting. The conservation of momentum is trickier. Say that you are in a small canoe in a placid lake and you have lost the paddle. You apply "body English" to the canoe, leaning forward gently and leaning back with force. The canoe begins to move until you retrieve the paddle. Children do it to get sleds started. It works *as long as you have friction.* If there were no friction between the canoe and the water, the canoe would move back just as gently as you moved forward, both of you taking your time. Then, when *you* leaned back forcefully, the canoe would move forward just as forcefully, to return to the exact position you were in before.

Energy is not being conserved. You are panting from the effort, using up oxygen, and developing an appetite. The law involved is the law of conservation of momentum.

John Bernoulli's problem of the shortest route, Huygen's study of the least length, Fermat's principle of least time, and the principle of least action of Maupertuis, Euler, Lagrange, and Hamilton all declare the inherent economy of nature in its cold, puritanical physical aspect. On the other hand, when nature is having fun, such as in the reproduction of life, it is downright generous, even prodigal.

We will see in Chapter 7 how the parsimonious universe of inorganic matter is running down, cooling off, while the generous universe of organic matter moves from disordered atoms to ordered molecules, from the cold slime of the earth to the movement of warm living things, in absolute defiance of the threat of ultimate cold death.

This shift from gloomy to hopeful thinking brings us back to Edmund Burke's statement that true economy is not parsimony. Conserving isn't enough. We have to do something with what we conserve. But we *do* have to conserve.

WHAT DOES IT COST?

This is a question calling for a simple quantity. All it tells us is that the cost is large, middling, or small. These are matters of fact. If we were to say it is *too* large or *too* small, we would be entering the field of value because value relates what *is* to what *should be.*

The question, "What *should* it cost?" was the step which led Lawrence D. Miles from fact to value. The question was asked within the framework of change for the better—exchange value!

Exchange, or trade, to be sustained and profitable, must be good for both parties. By exchanging firewood for game the freezing hunter and hungry woodcutter described in Chapter 3 eat roast venison near a warm fire. The proportions of the exchange determine whether this commerce will be profitable to both parties. This is one definition of a fair price.

Now suppose there are several freezing hunters and several hungry woodcutters. Each will offer different amounts of his product in exchange for what he needs, depending on his wants and resources. Before the end of the season some woodcutters will go hungry and some hunters will be mighty cold.

The name of the game is survival.

Comparing costs is as essential to survival in business as it is in the rest of the jungle.

Simple costs are matters of fact; comparing them with what they buy us, and with other costs, which provide the same utility, becomes a matter of value.

A Matter of Fact—Cost Visibility

To see product costs in perspective we must stand back and look at cost to the customer—that all-encompassing figure which includes the supplier's cost and profit, plus what the customer must pay others—such as repairmen—for use of the product, plus direct costs to him, such as price and the cost of investigation, appraisal, comparison, ordering, and financing.

This is a good time to review our own costs, as developed in Figure 5-2. Such manufacturing and administrative costs are carefully recorded, monitored, and compared in industry. Often whole departments and diverse programs exist for the purpose of controlling and reducing these costs.

Reducing cost to the customer is another story. As noted earlier, truly informative advertising, convenient displays, and good service under warranty constitute investments on the part of the marketing function which reduce those direct costs incurred by the customer over and above the sale price.

Certain other customer costs are particularly sensitive to the efforts of a value-analysis task group. Here are some examples:

COST OF OPERATION. Sensitive to reduced inputs for the same output. Examples are longer cruising range in an aircraft or longer battery life in portable equipment.

COST OF MAINTENANCE. Sensitive to improved lubrication, better gaskets, and better wearing surfaces in mechanical equipment, to longer periods between calibration of test equipment, and to more durable finishes.

COST OF REPAIR AND OVERHAUL. Sensitive to improved accessibility of replaceable parts and to ease of refinishing.

COST OF DOWNTIME. Sensitive to investment in better materials and components to reduce malfunctions.

COST OF DISPOSAL. Sensitive to planned subsequent use, such as packing cases convertible to workshop shelves or liquid containers shaped like canteens or Jerry cans.

A Matter of Value—Cost Comparison

Bearing in mind that in research, design, and manufacture the purchase of some materials and services is invariably required, the engineering and manufacturing members of the value task group must think of themselves as customers and must compare costs from a customer viewpoint.

The following units for cost comparison have proved useful in value analysis:

COST PER YEAR. This is the unit cost in quantities covering 12 months' production—if we are manufacturing—or 12 months' requirements—if we are buying. Unless otherwise directed, this is the basis of calculation for value-analysis savings.

COST PER UNIT OF OUTPUT. Examples are cost per watt, cost per Btu, and cost per horsepower.

COST PER UNIT OF MEASURE. Examples are cost per pound, cost per lineal inch, cost per square inch, cost per cubic foot, and cost per mile.

COST PER UNIT OF SERVICE. Examples are cost per passenger mile, cost per semester hour, cost per bit of information.

WHAT SHOULD IT COST?

Now that we know what it costs to perform the function—using the present approach—we need to answer the inspired question of Larry Miles, "What *should* it cost?" We usually seek a lower cost to perform the function, but we may come up with the same cost—better distributed —or with a greater cost that yields even greater usefulness. In any case, we should know where the money is going and what it is buying (see Figure 6-4).

The first answer we need is the BASIC or minimum cost of performing the PRIMARY function, the function assigned first-degree importance on the function definition chart in Figure 6-1.

ITEM	FUNCTION		DEGREE	PRESENT COST TO PERFORM FUNCTION	LOWEST COST TO PERFORM FUNCTION	RATIO
	Verb	Noun				
Present Item:						
REMARKS			Totals			

ITEM	FUNCTION		DEGREE	PRESENT COST TO PERFORM FUNCTION	LOWEST COST TO PERFORM FUNCTION	RATIO
	Verb	Noun				
Proposed Item:						
REMARKS			Totals			

Figure 6-4 What should it cost?

159

In the example mentioned under *Define the Function* we found that the function was to "bend water."

It was costing 23 cents to put the bend in the hose because a steel mandrel had to be inserted, the hose baked in an oven, and the mandrel withdrawn.

How to find the least cost of bending water? In this case the task group sent a purchasing trainee to the most economical stores in the area with instructions to buy the cheapest gadget he could get to "bend water."

The young man returned with a plumbing "U" which retailed for seven cents and could be bought in quantity for five. This is what bending water should cost. Note where it is entered on the example in Figure 6-5.

The marketing member of the task group protested. "Too hard," he said. "It may crack the laundry tubs."

A plastic one was found for three cents more. It had to have a shoulder to make it look decent, and that cost another three cents. The proposed plastic hose-termination would now cost 11 cents. That is more than the five-cent cost of the primary function, but a lot less than the 23 cents of the original method. Fully as important, the task group knew where the money was going and why.

At this state all we need to do is find out what it costs to perform the function, what it costs to provide the required benefits, and what is a good target for these costs. The target usually falls somewhere between the present actual, or estimated, cost and the lowest possible cost, as shown in Figure 6-5.

These charts, however, should be filled in only when doing so will be really useful. The charts may serve as no more than a framework for discussion; they are meant as aids to the job plan and not as ends in themselves.

FUNCTIONS IN THE TRANSPORTATION AND PROCESS INDUSTRIES

When the expenditure being considered is primarily that of a single resource, such as coal, the units of cost may be entered directly in units of that resource. The concept of resource availability can be extended to such limiting factors as weight, bulk, elapsed time or whatever penalty must be incurred to obtain the required result. The cost then may be entered in units of the limiting factor, such as the weight of fuel that a steamship must carry to complete a given voyage. The weight of fuel

| ITEM | FUNCTION | | DEGREE | PRESENT COST TO PERFORM FUNCTION | LOWEST COST TO PERFORM FUNCTION | RATIO |
	Verb	Noun				
Present Item: BEND IN DISPOSAL HOSE	REDIRECT	WATER	1	0.23		
	ABSORB	SHOCK	2			
	ENHANCE	APPEARANCE	2	DOES NOT DO IT		
REMARKS AS THE TEAM WROTE IT			Totals	0.23	0.5	4.6 to 1
AS THEY ACTUALLY DISCUSSED IT						
Proposed Item: PLASTIC TERMINATION	BEND	WATER	1	0.05		
		SOFT	2	0.03		
		PRETTY	2	0.03		
REMARKS			Totals	0.11	0.5	2.2 to 1
5 CENTS TO BEND WATER, 3 CENTS FOR SOFT, 3 FOR PRETTY = 0.11 ¢						

Figure 6-5 Example: What should it cost?

161

displaces payload and is an element of cost. Cubic feet of space occupied by the life-sustaining equipment in a space capsule is an element of cost because it limits the payload.

By and large, however, cost enters the formula for value in units of currency. The term for value, then, is expressed in units of performance per unit of currency. Say the function is to travel from Philadelphia to Chicago, first class, by air:

$$\frac{\text{function}}{\text{cost}} = \text{value} \qquad \frac{666 \text{ air miles}}{\$54 \text{ air fare}} = 12.33 \text{ miles}/\$1.$$

The purpose of the formula is to tell us what we get for our dollar, so that we can compare relative values. Traveling from Philadelphia to Los Angeles gives us more mileage per dollar, 14.42/$1.00, whereas from Philadelphia to New York we get only 5.8/$1.00. We were already aware that long-range air travel is more economical than short hops. Now we know *how much* more.

The formula follows all the laws of transposition:

$$\text{function} = (\text{value}) \, (\text{cost}), \qquad 666.00 = (12.33) \, (54),$$

$$\text{cost} = \frac{\text{function}}{\text{value}}, \qquad 54 = \frac{666}{12.33},$$

$$\text{value} = \frac{\text{function}}{\text{cost}}, \qquad \$12.33 = \frac{12.33}{\$1.00}.$$

Note that the unit of value, miles per dollar, is a compound unit made up of two dimensions: performance and cost. Sometimes it may include more dimensions, when performance itself is given in compound units such as miles per hour. In that case the unit of value would be miles per hour per dollar in three dimensions, distance, time, and cost.

The following example from the process industry illustrates how such compound units must be handled:

A continuous process treats 600,000 gallons of product in a 24-hour day at a daily cost of $12,000. Both the industrial efficiency and the economic value of the process are subject to a measure that answers the question, "What do we get for our dollar?"

We enter the measure of performance (process 600,000 gal/day), and the measure of cost ($12,000) in the formula for value:

$$\frac{\text{the function is process 600,000 gal/day}}{\text{the cost is \$12,000}} = \text{value}.$$

A simple equation? Not at all. It is an equality of *ratios* made up of two statements of proportionality, each using a different combination of units.

The first relates gallons to dollars:

$$\frac{600,000 \text{ gal}}{\$12,000} = 50 \text{ gal}/\$1.00;$$

that is, we process 50 units of product per dollar.

When we pay for the processing effort, we also pay for processing time; so we have a second statement which relates dollars to units of time:

$$\frac{600,000 \text{ gal/day}}{\$12,000} = 50 \text{ gal/}0.002 \text{ hr/}\$1.00$$

UNITS OF PRODUCT PER DOLLAR	UNITS OF TIME PER DOLLAR	UNITS OF COST IN DOLLARS
A = 50 gal	B = 0.002 hr	C = $1.00

To find TIME AND COST for a given QUANTITY, divide by A and multiply by B and C

1000 gal/A = 20   20 × B = 0.04 hr   20 × C = $20

1000 gal – – – – – – – – –each 0.04 hr– – – – – –at a cost of $20

25,000 gal/A = 500   500 × B = 1.0 hr   500 × C = $500

25,000 gal– – – – – – –each hour – – – – – – – –at a cost of $500

To find QUANTITY AND COST for a given TIME divide by B and multiply by A and C

1 min = 0.0166 hr
0.0166 hr/B = 8.3

8.3 × A = 415 gal   8.3 × C = $8.30

415 gal each minute at a cost of $8.30

8 hr/B = 4000

4000 × A = 200,000 gal   4000 × C = $4000

200,000 gal each 8-hr shift at a cost of **$4000**

Figure 6-6   Computing value in compound units.

$$\frac{24 \text{ hr}}{\$12,000} = 0.002 \text{ hr}/\$1.00.$$

The time and effort we get for our dollar is then

$$50 \text{ gal}/0.002 \text{ hr}/\$1.00.$$

As long as we keep track of the three types of unit—units of cost, units of quantity, and units of time—we can use this single expression to give us QUANTITY AND TIME for a given cost, as shown at the top of Figure 6-6, TIME AND COST for a given quantity, and finally QUANTITY AND COST for a given time interval.

DEFINING THE FUNCTIONS OF STATE DOCUMENTS

When in the course of human events it becomes necessary for one people to dissolve the political bands which have connected them with another . . .

Dissolve political bands was the necessary function that made it possible for the colonies to be free and independent states, but although the Continental Army and Navy had ensured independence it would take laws to secure freedom, so . . .

We the people of the United States, in order to form a more perfect Union, insure domestic tranquility, provide for the common defense, promote the general welfare, and secure the blessings of Liberty to ourselves and our posterity, do ordain and establish this constitution for the United States of America.

VII

CREATIVE PHASE

Having marshaled the information and determined what the product does, we ask the questions, "What else will do?" and "What will do it better?" which lead us to the creative phase of the job plan.

This chapter describes a creative session in a value-analysis workshop. It can also serve as a pattern for conducting creative sessions with engineering personnel or with a new product development group. The principles can be applied to individual as well as group creativity. As a matter of fact, well-conducted experiments, reported from time to time in the *Journal of Creative Behavior*, support the thesis that new ideas germinate most often when the individual is alone, but such ideas are based only on one individual's personal experience and personal store of knowledge.

In a creative session that includes diverse specialists task group members not only bring together a rich store of information but they provide stimulating discussion and a creative environment in which the seeds for new ideas are planted.

HOW TO WELCOME NEW IDEAS

The first step in the creative phase of a value workshop is to provide an atmosphere that encourages and welcomes new ideas. Such a welcome requires deliberate effort, for if an idea is really different and strikingly novel it is often met with hostility.

Why?

What is wrong with a departure from the ordinary and the expected? J. P. Guilford suggests that a "failure to match" between what we know and the new idea is a call to action. Either our view of a safe universe is wrong or something has happened out there that could mean danger. One way or the other, it is not good. We are not too different from those ancient kings who killed the bearers of bad news. If the new idea implies that either we are wrong about the universe or that we have to mobilize to cope with change, it is easier to kill the messenger than to adjust to changed conditions; so we try to destroy this upsetting concept by saying it won't work and we explain *why* it won't work. If we don't know why, we appeal to heaven with such pleas as, "If God had meant us to fly, he would have given us wings."

Normally subdued in the inventor, this tendency to fight new ideas can be suppressed altogether in a value task group. The members simply agree not to pounce on and destroy any new idea, not even their own. Baby ideas are then treated like babies and puppies—given a break until they are old enough to spank.

Criticism Postponed

Agreeing to treat new ideas gently does not make them safe. They are at least a potential threat to our comfortable view of the universe. If it turns out that witches *do* exist, be they ever so sweet-looking in their teens, they are nevertheless a menace to our unbewitched universe. Do we have to fight the entire new idea? What is wrong with isolating what is wrong with it and fighting only that or, better still, fixing it?

Criticism Transformed

Don Quixote de la Mancha was the noblest and most imaginative of all knights created by the mind of man. How would Don Quixote handle criticism of a proposed course of action? Let us say he rode into an enchanted forest where a fair maid was chained to every tree. Don Quixote believed that freedom is the greatest gift that heaven can bestow on man, and he occasionally helped heaven by breaking the chains of the enslaved—always against the advice of his down-to-earth esquire Sancho Panza.

We can hear Sancho's wail: "Don't free them, Sir Knight. One of them may be a witch!"

"Nonsense, Sancho," roars Don Quixote, "The world belongs to the bold. *You* find the witch and I'll free the others."

Presumably, Sancho gallops back to the inn to find an attractive broom suitable for testing witches. His objection is transformed into constructive action.

That is what the task group must do with menacing ideas—transform criticism into possible solutions. Instead of instantly presenting our objection, let us present its solution!

After we have a large selection of potentially good ideas we can judge them critically. In the meantime, we postpone critical judgment. Instead of saying, "It will not work," we can look for reasons to make it work.

I had the pleasure of being one of the instructors at a London seminar directed by Fred Sherwin, past president of the Society of American Value Engineers. As we entered the creative phase, Fred Sherwin reminded the group that they were to postpone critical judgment, that when an idea was presented nobody should say, "It won't work."

A tall Yorkshireman growled, "It won't work!"

"But nobody has come up with a single idea yet," I protested. *"What* won't work?"

"Inventing on schedule," the tall man rumbled, "That's what won't work!"

There is something sinister about the way certain words are pronounced in Yorkshire. "Moreover," the tall man added, "I want to know why we can't say 'it won't work.' "

I thought we had explained that. The bit about new ideas being frail and easily crushed, but my buddy Fred Sherwin knows that there is a time for persuasion and a time for compulsion. "Because," he snapped, "Anyone who says, 'It won't work,' will have to put a half-crown in this bowl—"

Stalking toward the bowl, the Yorkshireman produced his half-crown, but Sherwin continued, "—we will send the money overseas to the Ton-Ton Macous."

Now, nobody knows who or what the Ton-Ton Macous really are, but they are certainly un-English. The tall man put the money back in his pocket. "But you have to show me," he challenged. "Why don't you invent, *on schedule,* say a better mousetrap, within a quarter hour."

Ideas began pouring fourth. Nice, safe, time-tried ideas: cat, poison, etc.

I pointed out that we needed many, many ideas; hopefully, some way-out ideas. If it was obvious that they wouldn't work, we should think of a way to *make* them work.

Our Yorkshireman rumbled, "I have a way-under idea, a unidirectional tunnel leading to a neighbor's house."

A scholarly type—tweeds, pipe, horn-rimmed glasses—took out a half-crown. Later he explained to us that he was thinking about a concentration of complaints, possibly backed by firearms, coming from the neighbor's house. How to fix it? Pocketing the coin, he gestured in all directions, "Random distribution catapult; everybody hates you just a bit, but nobody hates you enough to shoot you," and looking anxiously at the instructors, "It *is* a step forward, isn't it?"

Before he had actually finished a bushy-browed Scotsman growled, "Castr-r-rate the little beasts!"

There was a long pause. Not too many people have seriously considered castrating mice. Then a heretofore quiet member of the audience slowly rose to his feet, his inspired features bathed by a beam of light from an upper window. "I've got it," he announced, in mystical rapture, "birth-control cheese . . . !"

In Stockholm it was, "Make friends with the mice," and "Teach them to do something useful," but, just as the British in London, the Norwegians in Oslo, and the Danes in Copenhagen, the Swedes also homed-in on birth-control cheese.

In Paris it was different. "Birth-control cheese! Mon Dieu! A typical Anglo-Saxon repressive measure which nevertheless leaves you with one generation of lonely mice." This was a computer design engineer. "Now, you in North America," he added, "have a weed hormone which you spread upon the grass around your houses and the weeds such as the lion dandies—"

"The dandelions?"

"Yes, the dandelions and other weeds of broad leaves begin to lead a riotous existence. They grow themselves to death and collapse in happiness!"

Before I could question his botany, he suggested, "Now, add to the cheese a sex hormone and both Papa Mouse and Mamma Mouse die under the most pleasant of circumstances."

This Frenchman knew his value analysis. "The function of the mouse trap," he continued, "is not to do injury to the mice but to eliminate them as painlessly as possible, no?"

"Eliminate them, yes," I agreed, "but pain to the mouse is of little concern to the purchaser of mouse traps."

He nodded, "Yes, the purchaser may be stupid."

"Why stupid?"

"Monsieur Fallón, one of the books you lent me shows that force, compulsion, and hostility is not the best way to do anything. The more unpleasant you make the operation for the mice, the more ingenious they will become to avoid it."

I had lent him two books, William J. J. Gordon's *Synectics* and *The Rational Manager* by Kepner and Tregoe.

"From *Synectics*," he continued, "I learned also about personal analogy. Were I a mouse, I inquire of myself, what would reduce my ingenuity and make me march to death voluntarily? What makes men face, with their back, the bullets of husbands? Sex! Sex can take mice from the cruel guillotine of the mouse trap to an idyllic death in a moment of romance." *Naturellement*, they go more willingly.

"Did you get a chance to look at the other book?" I asked him.

He smiled. "But of course! It was in *The Rational Manager* that I discovered how your clients arrived at birth-control cheese. You use the book in your work, no?"

"Yes, I use it, and *Synectics*, in all my workshops."

"And you see about the birth-control cheese?

"No."

"Messieurs Kepner and Tregoe," he reminded me, "insist on finding the cause of a problem. Mice are a problem. Big mice come from little mice. Little mice are born. Birth-control cheese cuts off the problem before it is born."

He was right, and the books *are* good.

FREEING YOUR INVENTIVE PERSONALITY

The most recent investigations (1970) indicate that creative potential is universal among human beings. Although there is wide variation, everybody is inventive to some degree—at least as inventive as the chimpanzee who can join two sticks together to reach a banana. There is an inventive personality within the least imaginative reader of this paragraph, and certainly within all the members of a value task group. Their industrial activities, in contributing value, *require* inventiveness.

What Makes a Creative Person?

Early studies of notably creative families (Galton, 1869) seem to indicate that the genetic make-up of the individual—his creative heritage—is an important factor. The circumstance that a creative family provides not only creative genes but also a creative environment is often overlooked. A value task group cannot change the genes but it *can* modify the environment.

Value analysis works with people as they are; people as they are usually have more potential than they, their bosses, and their families, for that matter, might think. A value task group therefore contains ample inventive potential. How much of this potential becomes reality is a function of the creative environment provided by the workshop atmosphere.

ENVIRONMENT FOR INVENTION AND DISCOVERY

Many studies have been conducted on the extraordinary art of Michelangelo and the inventiveness of Leonardo da Vinci. One such study by Gilbert (1967) goes a little deeper. It asks *why* and goes on to describe the creative environment provided by the house of Medici, the city of Florence, and the families and patrons of the two men.

Florence, like the Athens of Phidias, the Persia of Omar Khayyam, and the Cambridge of Newton, provided a climate for artistic creation and scientific discovery. Omar was as good a mathematician as he was a poet and philosopher. One wonders how well he, Michelangelo, and Newton would have done had they lived in Sparta, in the Spain of Torquemada, or in Cotton Mather's New England.

Before they set out to create anything else members of a value task group must therefore generate an environment in which latent inventive traits rise to the surface. Essential for this condition is an atmosphere of freedom from reprisal, freedom from ridicule, and freedom from the weight of tradition. Such an atmosphere engenders the realization that rules were made for man, not man for the rules, and that the laws of nature are laws only until new laws are discovered.

FOSTERING INVENTIVE TRAITS

Here are some of the traits, latent in everyone of us, that the environment of the workshop can bring out.

RECEPTIVITY OR OPENNESS TO NEW IDEAS. As Professor Alamshah puts it, the glasses should be ready to receive the wine before we open a bottle of champagne.

VENTURESOMENESS. Together with honor and the love of liberty, *el espiritu de aventura* was Don Quixote's guiding light, and a very profitable light it turned out to be. After 300 years the book is still a best seller. Potential profit distinguishes a venture from a simple act of boldness.

One source of profit in the task group's venture is the personal gain of the members as they learn to work efficiently and harmoniously with people whose skills differ widely from their own; another is the revelation that they can get better-than-usual results from other-than-usual methods. The group, as a group, becomes a joint venture that encourages and supports daring ideas.

DISCERNMENT. As a group trait discernment can be developed by jointly identifying the few essential elements in a pattern of information and by recognizing those gaps that weaken the pattern.

CURIOSITY. This trait, as developed within the task group, means that the information phase is never really over. The interacting curiosity among members leads to the recurring question, "What would happen if . . . ?" The answers follow two different patterns of thought:

CONVERGENT THINKING. A trip into the sunlit countryside of well-organized knowledge. Rich in tradition and experience, this serene land has good roads and solid bridges. It is the land that provides the resources, the scientific background, and the jumping-off place for the expeditions that venture into . . .

DIVERGENT THINKING. The unexplored, uncharted domain of the unknown; a mysterious new world where mountain passes must be discovered, torrents bridged, and pitfalls sensed and avoided. This strange universe of mirages, phantoms, and auroras offers no advance information but much oportunity. Each treasure discovered leads to another and still another, and all the wealth returns to the mother country from which the expedition set forth.

Convergent thinking sets objectives, tells us what is worth searching for, and provides a point of departure and a path of return.

Divergent thinking generates way-out ideas, ignores road maps, KEEP OUT signs, and the dire warnings and headshaking of those who live by the old rules.

Most writers on creativity emphasize divergent thinking at the expense of convergent thinking, but Thomas S. Kuhn, whose field is the history of science, suggests that the evidence of creative production throughout the ages points to an interplay such as I have just described. Professor Kuhn says, "we must seek to understand how these two superficially discordant modes of problem solving can be reconciled both within the individual and within the group."

In a value task group we do just that. The fuel for convergent thinking is implicit in the very composition of the group—a combination of traditional skills that marshals accepted knowledge. From this solid ground the group sets forth on the adventure of divergent thinking. The trick is to keep the habit of looking for signs and bridges from making us overlook promising trails and feasible river crossings. We still have to exercise enough control to make exploration really fruitful. This gentle control is sometimes called *self-limitation*.

FLUENCY. The ability to generate many ideas in a given time. An individual's fluency of thinking, of association, and of speech may well be fixed. The basic capability is not likely to improve in the few days of a value workshop, but the motivation to express and explain whatever ideas do germinate is enhanced by the spirit of investigation, consultation, and discussion within each task group.

FLEXIBILITY. The ability to change direction in search and thought. Here we have an example of what Kuhn calls "the essential tension," the conflict between the world as we think it is and the world as it actually turns out to be.

An Amazon Indian launches his canoe and paddles to intercept the braid of ripples raised by a giant fish, the barely edible pirarucu. He works his way to the bow, raises a harpoon, and suddenly sees that a hundred pecaries—delicious little pigs—are swimming across the river. Dropping to his knees, he grabs a paddle, overtakes the pigs, harpoons the leader, cuts its throat, removes the harpoon iron, and quickly puts it back on the shaft. Then, he paddles to overtake the disorganized herd and get a few more pecaries. That's flexibility!

One of us would have said, "Harpoons are for fish, not for pigs"; and, as both fish and pigs disappear, "My degree is in fishing, not hunting."

TOLERANCE. Unperturbed acceptance of the ambiguous, the absurd, and the paradoxical. "So the pink steamroller gives milk? . . . So what? So it's wearing a bikini! . . . ? Must be a dream. Well, then, if it's a dream, let's put Raquel Welch in the bikini. Ah . . . that's better." This process is called *transformation*. The example was chosen because it was easier to explain than the equally puzzling behavior of electrons within a crystal lattice when they are energized by high-speed photons; something a creative physicist must learn to accept with equanimity.

Ingredients of a Creative Session

The stimulant is animated discussion in an atmosphere of economic safety and intellectual freedom. A wide diversity of skills provides the fuel. Dormant or suppressed inventive traits suddenly blossom forth, popping explosively out of their buds, scattering pollen in all directions, then remaining open and receptive to be pollenized in turn.

Poetic? . . . Nonsense! It's sexy and it leads to serendipity (Moore) which is sheer gravy—a golden platter of happy, unexpected, outcomes.

Summary of Inventive Traits

Receptivity	Divergent thinking
Venturesomeness	Fluency
Discernment	Flexibility
Convergent thinking	Tolerance

THE PROCESS OF INVENTION AND DISCOVERY

Jules Henri Poincaré, cousin of Raymond Poincaré, the statesman, is recognized as the most brilliant mathematician at the turn of the century. Called "the last of the universalists," he developed the celestial mechanics we use today for directing the orbital movements of spacecraft and for docking satellites. He understood and advanced creative work on all branches of mathematics known in his day, being one of the first to make extensive use of non-Euclidian geometry in mathematical physics.

The Psychological Society of Paris asked him why he was so inventive. He answered that he was not—at least not more than anyone else.

The conversation must have gone something like this:

"Oh, yes you are! Your colleagues tell us that you have invented new mathematical tools and have discovered new mathematical entities."

"Such as?"

"The Fuschian functions!"

"Oh, that," replies Poincaré, "Well, I'll tell you," he confides, "those Fuschian functions gave me a lot of trouble. I worked, and worked, and then . . ."

But you can read Poincaré's own words in his *Science and Method*. Hadamard makes a good analysis of Poincaré's approach.

The "Classic" Steps—Preparation, Incubation, Illumination, and Verification

We owe this picturesque description of the inventive process primarily to Poincaré and Wallas. Poincaré described how the long and hard work of tracking down an idea was followed by a period in which the search was put aside—either purposely or for other reasons. He labeled this period *incubation* and assumed that the "unconscious or subliminal self" goes to work, combining and recombining all inputs received during conscious effort and permitting only the interesting ones to enter the realm of consciousness.

Is it simple chance that confers this privilege? Evidently not, according to Poincaré. Only those stimuli that deeply affect our emotional sensibility are privileged to merge into the conscious. Emotion, he maintained, *has* a place in mathematics; otherwise we would not talk of mathematical beauty. His own words are freely translated.

To what mathematical concept do we attribute this character of beauty and elegance which produces an aesthetic emotion? To those elements, so harmoniously arranged that the mind can embrace the whole and still grasp the details.

Such concepts do not, as a rule, filter piecemeal to the surface, but like Athena they burst forth, fully armed, in what Poincaré called

. . . a moment of illumination, always brief, always sudden, and always accompanied by a feeling of absolute certainty.

Because this "feeling of absolute certainty" can play tricks on us, Poincaré called for a period of *verification*. After all, an idea may be promising because it looks good, but this does not mean that it will work.

It is to the credit of Poincaré and Wallas that the creative steps identified by them still serve as a foundation for the work of many modern writers on invention and discovery. Perhaps the only substantial modifications to the classical steps is the experimental observation of Catherine Patrick that the steps do not always follow in 1, 2, 3 order. J. P. Guilford, after carefully studying these classical steps, offers an operational model for problem solving in general. The reader, who has followed me this far, should read about this model in the context of Guilford's excellent book. Here is a brief description of the classical steps, with the terminology updated when necessary:

PREPARATION. This is a very broad term covering the following tasks:
Identifying the wants to be satisfied
Gathering existing information
Searching for additional information
Random exploring
Appraising the information for relevance and truth
Analyzing the resulting knowledge for usefulness

The preparation stage ends, hopefully, in a planned interruption or recess (more often in fatigue or frustration and entirely too often in lack of time for further preparation).

WITHDRAWAL. The "incubation" stage of Poincaré and Wallas. A modern objection to the term *incubation* is that we do not know enough about what happens at this stage to call it by that term. *Plateau* has been suggested, but that also implies greater knowledge of the creative process than we have now. *Withdrawal,* on the other hand, is something that we do or something that happens to us. It does not lend itself to the complaint, "But sir, I don't know how to incubate" or "How can I reach a plateau when I am not getting anywhere to begin with?"

INSIGHT. The classical "illumination" stage. Because of its diverse sources of word roots the English language often has an exact word for a concept which "purer" languages have to describe by analogy. *Insight* is an exact English word for what other languages must compare with "illumination."

VERIFICATION. Checking it out. This is where the classical and modern terminology converge. The term needs no explanation.

The stages usually associated with the process of invention and discovery are therefore

PREPARATION
WITHDRAWAL
INSIGHT
VERIFICATION

Although the value-analysis job plan starts earlier and follows through to implementation, it does full justice to the process of invention and discovery, as shown in Table 7-1.

The Controversy on Brainstorming

A team of surgeons planning an operation, a military staff planning a campaign, or a jury arriving at a verdict are seldom accused of "potluck group-think." The result of their endeavors is not considered "cerebral popcorn"; yet these terms have been used to describe Alex Osborn's *brainstorming* (Benson). Less picturesque but more significant is the faint praise of Kepner and Tregoe: "This technique does sometimes produce a novel way out of a problem situation."

The article and the book cited are only a sample of the hostility that one often encounters toward "brainstorming" in particular and "creative problem solving" in general.

Why?

A quite understandable reason is that, despite its overpersuasive approach, Osborn's method works so well that it creates enthusiasts whose dramatic presentations occasionally get in people's hair.

The identical thing happened in Anatole France's *Penguin Island.*

Table 7-1   The Creative Process and the Value-Analysis Job Plan

Creative Process	Value-Analysis Job Plan
	Information Phase
PREPARATION	Define the general objective and specific goals
	Gather existing information
	Search for additional information
	Explore at random for surprise information
	Appraise information for relevance, freshness, and accuracy
	Analytic Phase
	Analyze resulting knowledge in the light of the product's function
	Relate the elements of product worth to their corresponding elements of product cost
	Shift thinking from matters of fact to matters of value, from what *is* to what *should be*
	Creative Phase
	Reinterpret, transform, and rearrange data in order to generate new ideas
WITHDRAWAL	Give ideas time to mature by going over the data that may affect them
INSIGHT	Develop better methods of performing the product's function, a better function to satisfy the customer, or eliminate the need for the function
	Convert ideas into concrete options
	Evaluation
VERIFICATION	Verify technical feasibility, customer appeal, savings in time, savings in dollars
	Relate benefits to their cost
	Estimate cost to implement, time to implement, payoff period
	Appraise risk
	Compare the various options
	Presentation
COMMUNICATION	Describe choices, forecast consequences, recommend action
	Implementation
	Present plan of implementation, obtain approvals, assign work, follow up, and report progress

The converts became a problem. Converted penguins, untainted by original sin and untrained to sin at all, were unbearably virtuous and even the angels resented them. In the same way Osborn's convincing writing in *Creative Imagination* may have been too inspiring for those readers who, having great spiritual resources and a vocation for the clergy, adopted Creativity as a religion.

As noted earlier, neither value analysis nor group dynamics has escaped the well-meaning enthusiasm that gives rise to cultism. But now we are considering this problem as it affects our use of Osborn's method. Are we to lose the benefits of the method because we find the terminology distasteful? Because showmanship wounds our scientific sensitivity?

Of course not, we are rational human beings; there must be something *else* wrong with brainstorming. Surely anything described by such a . . . such a Madison Avenue name *must* be unscientific.

So we search the literature—the publications of learned societies in the fields of psychology and education—and we find that while the detractors were detracting and the defenders were defending Alex Osborn, not a cultist himself, was most scientifically modifying, improving, and advancing his method in such a way that both detractors and defenders were left behind. Every revision of his book contains an improvement, and many of the improvements are based on experimental findings originally thought to disprove his claims.

As the dust of controversy settles, the most important principle of brainstorming, *suspending critical judgment,* emerges as an accepted way to stimulate and increase the flow of ideas.

Osborn's method, as finally described by him, takes into account those experiments in which the subjects, as individuals, generated more ideas than they did as a group, *provided they postponed critical judgment.*

In value analysis, however, group interaction is essential. We do not organize *random* subjects into groups so that they may innovate. Instead, we carefully select a mixture of specific skills in order to integrate the components of product value. Such a group has a far richer pool of information than any one of its members. The purpose of *our* group action is primarily the interchange, comparison, and validation of relevant data.

CREATIVE PHASE OF THE VALUE-ANALYSIS JOB PLAN

Once all members have acquired the knowledge developed by the group they move from matters of fact into matters of value, from what *is* to what *should be.* Quite often—although not always—*what should be* has

to be invented. Ideas develop to meet the specific needs or desires identified by the group. Such ideas may arise then and there or they may
occur to individual members during a trip to a supplier's plant, during
lunch, at home, or while driving to work next morning.

When an idea develops, you have to be nice to it. You cannot afford
to discourage the shy little one with web feet and gills—it might turn
into a dragon just when you need a dragon. If nothing else, you will have
a lot of interesting friends in the kingdom of the mind, and, believe me,
they will turn up when you need them most.

THE TASK OF INNOVATION

Change is one of the basic facts of life. It can create problems and solve
problems. It can create opportunities or cause us to miss opportunities.
The tactics for coping with change are

ignore change and let it happen to us
anticipate change and adapt to it
create change and profit from it

The task of updating, previously described, constitutes a conscious and
systematic effort to apply the second tactical approach. It represents an
active defense but only a defense. To take the initiative a value task
group must undertake the task of innovation, a planned and studied
group effort to free each member's ability for combining, modifying, or
discarding existing concepts, for creating new ones, and for putting these
concepts to work in one form or another.

The work of Poincaré, Wallas, and Hadamard has served as source
material for a number of excellent books published in the United States
on creation, invention, and discovery. The literature is well known
thanks to the Universities of Utah and Southern California and New York
State University at Buffalo, as well as the work of the Creative Education
Foundation at NYSU Buffalo.

When we ask the classical value-analysis question, "What else will do
the job?" we are only opening the door to innovation. To release the
innate inventiveness of task group members we have to remove barriers
to innovation, provide aids to novel thinking, outline the mental process
for advancing from the common to the uncommon, and conduct a
planned search on the one hand and some random exploring on the
other.

OUTLINE FOR THE TASK OF INNOVATION

Identify the Barriers to Innovation

Undue dependence on custom, tradition, procedures, etc.
Undue caution
Lethargy
Fixed ideas
Established patterns of behavior

Surmount the Barriers to Innovation

Abandon your regular patterns of thought
Momentarily ignore what you learned in school
Forget the way things are supposed to be done
Question tradition
Suspend critical judgment
Seek a great number of ideas
Seek a great variety of ideas
Appraise the cost and risk involved in *not* taking chances

Adapt to Change by Considering

New materials
New methods
New customer requirements

Create Change Through Variations

The means of satisfying wants: intermediate objectives, supposed
  benefits, inputs and outputs
The interaction of elements: circuit, cycle, flow network, linkages
The use of time: sequence, duration, frequency
The use of space: arrangement, orientation, form factors, size
The disposition of matter: state, density, distribution of mass, weight
The use of energy: sources of power, modes of drive

Change the Point of View

Identify with the product—personalize the problem
Stand back from the product—put the problem in perspective

Look for Analogies

In the life sciences
In the physical sciences
In your own industry
In other industries

Look for "Way Out" Ideas

List primitive or elementary methods
Examine the absurd and paradoxical
Forecast methods of the future

Examine and Screen the Choices Developed

Simulate or forecast consequences
Exercise critical judgment (deferred until now)

BLAST, THEN REFINE

Lawrence D. Miles originally used the expression *Blast, Create, Refine*
to describe the assault made on the shroud of tradition, inertia, and
sanctity that often obscures the function of a product. The technique is
delightfully presented by Miles in his 1961 book. He describes a pioneer
couple's three-room dwelling and outhouse, the various additions, and
the final structure that can accommodate a large family, plus indoor
plumbing. The *Blast, Create, Refine* technique blasts away all the patch-
work aspects of such a house, determines what the new functions should
be, and, finally, refines the design.

The reader will note that I have developed this concept in Chapter 6,
where we blast away the structure to determine the function, and in this
chapter, where we strive to create new ideas.

Perfecting and Simplifying New Ideas

One definition of the verb, *to perfect,* is to bring *to final form: to complete.* In order to complete our ideas, we must blast away the scaffolding
of the creative process, leaving only the final idea in terms of dynamic
satisfaction of wants: the function and its resulting benefits. Then we
must refine the idea by achieving the right balance and proportion
among the benefits. Finally, we must simplify the means of providing
those benefits.

THE PRINCIPLE OF BALANCE AND PROPORTION

Then let us mingle our ingredients, with a prayer to the gods, Dionysus or Hephaestus or whichever god has been assigned this function of mingling.

Plato in the *Philebus*

The amount of resources, such as materials and direct labor, that can go into an industrial product is determined by what the customer will pay, less the indirect expense and profit of the suppliers.

Given this limitation of resources, we know that we cannot increase one element of product worth without reducing one or several of the others. What we have to work with, then, are the proportions among the elements of worth that will yield the best value.

Blindly striving for the most of the best of everything amounts to trying real hard instead of thinking. Achieving a balanced combination of benefits, on the other hand, calls for common sense, not brute effort. "A painter," said Aristotle, "would not give his creation an oversize foot, be this ever so magnificent a foot." Neither can we endow our creations with exaggerated portions of any one benefit, whether it be economy, reliability, performance, or what have you. Instead, we must determine the relative importance of the desired benefits and combine them in the measure and proportion that will satisfy the customer, for, as Plato said in the Philebus, "the qualities of measure and proportion constitute both beauty and excellence."

Had not the classical Greeks understood the principle of balance and proportion better than any other people, I would not be quoting them here, but understand this principle they did, leaving ample evidence of its application in their sculpture, architecture, science, and statecraft.

In Book II of his *Politics*, Aristotle repeats, ". . . it is the perfect balance between its different parts that keeps a state in being." With respect to an industrial product we can say that, for a given level of resources, the balance and proportion among desired benefits determines the worth of the product.

THE TASK OF SIMPLIFICATION

In a comprehensive study of more than 100 value-engineering implemented recommendations the American Ordnance Association found that

21% improved performance
38% improved quality
39% reduced weight
40% simplified maintenance
46% made the product more reliable
76% reduced lead time
90% made the product easier to produce

Perhaps the least understood and most valuable of these improvements is the reduction in design, procurement, and manufacturing cycles which releases all manner of resources.

In Chapter 2 I described how a value task force was set up for the express purpose of reducing the cost of a production contract. The group not only reduced cost but came up with improved performance, higher reliability, and lighter weight.

Normally we do this on purpose, but performance, reliability, and weight were satisfactory in this equipment. The customer would not pay for improvements in any of them. Cost to us, however, was not satisfactory. The reader will recall how the group went to work looking for dollars alone, found the dollars, *and* an improved product!

Why did this task force, making no effort to improve product worth, nevertheless come up with a better product? They had simplified. Simplification means fewer parts, fewer interfaces, fewer contacts, fewer things to go wrong, lower resistance. The reduction of mechanical friction and electrical resistance results in lower power requirements, smaller transformers, smaller motors and, of course, lighter weight.

An egg is the acme of simplification. Compare it with a package of blueprints for building a chicken. This brings up a question. Why don't we turn out simple products to begin with? Why do we have to simplify?

A good point. But look at the picture of a 1920 airplane and an airplane today. Now look at a picture of a 1920 chicken. A man-made product changes and adapts much faster than natural organisms, which have taken hundreds of thousands of years to arrive at their present state of ordered, proportioned, and streamlined complexity.

Note that we are talking about *simplification,* not *simplicity.* Simplification, as we use it, means the elimination of excess complexity. The whole trend in natural organisms and human products is toward greater and greater complexity. This in turn generates more and more excess complexity to be removed.

Some of this excess complexity is the natural and inevitable result of step-by-step product development. This is part of the dynamic process of growth. Every step forward in product design obsoletes some aspect of the earlier design. The remaining physical part may not be worth re-moving at the time, but sooner or later there will be enough parts to justify planned simplification.

Even the best thought out designs require a systematic search for unnecessary complexity. There would be no new models unless the manufacturer and the customer jointly carried out this search. The customer seeks a product that will serve him in a more direct, more pleasing, and simpler manner, and the manufacturer searches for a way to give him just that.

The question is when to stop refining the product and to put it on the market. In making this decision we should remember that many an "overengineered" product is really underengineered—someone did not have time to simplify.

The Task

Chart shorter and straighter paths for conveying information, for con-verting and transfering energy, and for transmitting motion.

The Techniques

ELIMINATION
COMBINATION
REARRANGEMENT
MODIFICATION
SUBSTITUTION
STANDARDIZATION

VIII

EVALUATION

All the "keys to value" and all the "reasons for needless cost" hinge on good or bad decision patterns. The value specialist does not make decisions for others, but he must provide the kind of methods for scientific decision that can actually be used by the diverse specialists who control product value.

OPTION, CHOICE, SELECTION, AND DECISION

An *option* is defined as the right to exercise the power of choice. *Choice* implies the opportunity and privilege of choosing freely. It also imposes the requirement of giving up what is not chosen. The act of *selection* is defined as picking out one or more items from a larger number of similar or analogous items. This action is the same as rejecting the ones not selected. Choice and selection, therefore, imply a measure of conflict, but it is usually mild conflict *within* the decision maker. You pick out an apple and the other apples don't seem to mind.

Decision, on the other hand, settles external conflict of one sort or another. It *decides* the issue within the prize ring or in court. It also resolves the less evident conflicts among design objectives such as performance, reliability, low cost, and early delivery.

In the decision process much of the conflict centers around the objectives and their relative importance. Once this has been settled the task is reduced to comparing probabilities and information in order to make a simple choice or selection among the available courses of action.

NUMBERING, MEASURING, WEIGHING

If from any craft you subtract the elements of numbering, measuring, and weighing, the remainder will be almost negligible.

Plato in the Philebus

Plato divided the arts and crafts into two categories: the more scientific, such as architecture and shipbuilding, and those that depend on "rule of thumb, toilsome practice, and good guesses," such as medicine, farming, and strategy. He considered those that use numbering and measuring "immensely superior in accuracy and truth."

Plato did not know that while he was writing these very lines a mysterious professional soldier in China's period of The Warring States, the very Sun Tzu whom we met in Chapter 5, had successfully applied quantitative methods to warfare. This warrior-scholar had condensed into 13 brief chapters "the concentrated essence of wisdom in the conduct of warfare . . . never surpassed in comprehensiveness and depth of understanding," as B. H. Liddell Hart says in his foreword to Samuel B. Griffith's translation of *Sun Tzu, The Art of War*.

Liddell Hart is Field Marshal Montgomery's favorite military writer. Such praise from him adds a modern professional endorsement to Sun Tzu's words (after Griffith and Kuo Hua-Jo):

Now the elements of the art of war are: first, measurement of space; second, estimation of quantities; third, calculations; fourth, comparisons; fifth, probability of victory.

On the merits of quantitative methods Sun Tzu was explicit:

Quantities derive from measurement, numbers from quantities, comparisons from numbers, and victory from comparisons.

THE *COMBINEX* AID TO DECISION

To follow Sun Tzu's pithy instructions we determine the *quantities* that will contribute to attaining our objective, select a suitable scale of *measurement*, choose *numbers* that can be meaningfully compared with one another, and then provide a framework for *comparison*. . . .

The COMBINEX Aid to Decision consists of the following:

1. Analysis of the objective to identify its requirements or expected benefits.

2. A weighting technique based on the principles of balance and proportion and of limited resources.

3. A bounded interval scale which excludes both the inadequate and the excessive in order to measure variations within a practical range of choice.

4. A scale of commensurable ratings representing the contribution made by each of the available choices toward each of the requirements or benefits.

5. The COMBINEX scoreboard which serves as a submatrix for combining the benefits that make up a requirement and for noting their relative importance. Finally, the scoreboard provides a framework for evaluating various courses of action or choices in order to select

(a) the best combination of desirable characteristics for a given level of resources,
(b) a given combination of desirable characteristics for the least cost in resources, or
(c) the best relationship between desirable characteristics and their cost in resources.

More important, putting numbers into this framework makes people think. The scoreboard serves as a two-dimensional check list which forces us to face, one at a time, the major factors leading to a decision; for instance, at this point we are obliged to study the choice between the *best combination for the given resources* or the *least resources for a given combination.* Trying for both the best combination and the least resources is like trying to get both ends of a seesaw to go up at the same time. No technique of mathematical optimization can achieve such an effect.

Mathematics can work only with what we have to begin with, but other methods can be used to improve what we have. We can put a jack under the seesaw, and by throwing in additional resources we can then raise both ends of the seesaw at the same time. The COMBINEX method can point out those areas in which additional resources can be spent to greatest advantage. Where do these resources come from? From the elimination of waste, of course.

Example: A Battlefield Radar

We set up our scoreboard with the requirements along the top and the various choices in a column down the left. Here is a hypothetical ex-

ample: a transportable battlefield radar to be designed for sale to one of
the larger South American nations. Their needs include high *mobility*,
low initial cost, which we will call *initial economy*, low field cost, or
field economy, good performance, and a property which the customer
calls *combat endurance.* The last includes reliability, low repair time,
negligible maintenance, the capacity to withstand the shock of near
misses, and the capacity to withstand rugged transportation conditions.

The Art and Science of Combination

Once the customer's requirements have been defined, they must be
studied systematically in order to generate a number of satisfactory
combinations. The key concept here is that of a *balanced combination*
rather than the most of the best of everything. This is the difference
between well-allocated effort and maximum effort in all directions.

The most promising combinations are then compared by expressing,
in numerical measures, the extent to which each element of each combination contributes to the particular requirement it is supposed to meet.

When we suggest numerical measures, people frown. They look at
each other and someone says, "You can't compare peaches and pears!"
But you *can.* It is no problem if you take one property at a time. A pound
of peaches weighs exactly as much as a pound of pears. A carload of
melons occupies the same number of cubic feet as a carload of potatoes.
A dozen elephants is just *one* dozen and has the same number of units
as a dozen ping-pong balls. At this point someone will say, "Ha! Just
what property have *endurance, mobility,* and *economy* in common?"

Meaningful Numbers

Endurance, mobility, and economy all contribute to the mission of the
equipment. How does economy contribute to the mission? It contributes
numerically, for cost is the inverse measure of the *number* of equipments
you can get from the resources committed. Cost, size, and weight determine *how many* equipments can be provided in a given place at a given
time. This contribution, this satisfaction of the customer's needs, is the
common property that our rating scale measures for all major characteristics.

In school "reading, riting, and rithmetic" are measured numerically
on a standard scale in which 70 is passing, 90 is very good, and 100 is
perfect. This scale has the advantage that it is immediately meaningful
to everyone—engineer, shop foreman, buyer.

We can use this standard scale, but we may not want to pay the full

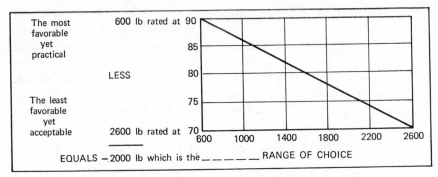

Figure 8-1   Rating the effect of system weight: a linear normalization.

price of perfection implied by the grade 100. Absolute perfection is the realm of the gods; when you get up there, they clobber you! It costs too much and takes too long. So the real end points of our scale are 70 and 90. Figure 8-1 shows how we can transform system weight into a rating scale. This figure is based on the assumption that the correlation is linear. Often, however, the relationship is not linear.

Effectiveness and Utility

As any marine engineer knows, the amount of fuel consumed is not directly proportional to shaft horsepower delivered. In the same manner, lighter weight is not directly proportional to improved mobility. In Figure 8-2 one can see that as objects get lighter and lighter they do not get uniformly more mobile. At first you can't budge them; then you can barely budge them and finally they begin to move. This nonlinear rela-

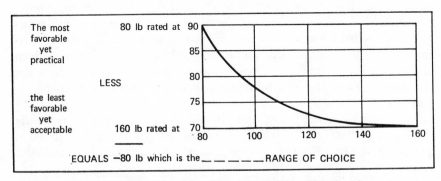

Figure 8-2   Effect of weight of heaviest case: the utility function.

tionship in physical interaction is a counterpart of the utility function in economics. Such nonlinearity should be suspected, searched for, and taken into account.

It seems easy enough to understand that an article can be worth more to one person than to another or that it can be worth more for one particular application than for another. Economic utility, or worth to a particular customer for a particular application, becomes increasingly important as the number of free individual customers increases. Daniel Bernoulli formulated a good mathematical approximation of utility as early as 1738 (Somer). He was followed by Herman Gossen, William Jevons, Maffeo Pantaleoni, and Alfred Marshall. John Von Neumann and Oskar Morgenstern combined the previous groundwork with their own innovations to synthesize a theory of utility applicable to present-day economics and to the modern analysis of value.

The customer's concept of value begins with his personal utility, which he quantifies by saying, "This is *worth* so much to me." The product must *do* something for him. He has to want it before he even considers the cost. Then he compares what the product is worth to him and what it costs him.

A Measurement of Customer Satisfaction

We try to give the customer either a little more for his money or something every bit as good for less money. The techniques of innovation, simplification, updating, and better use of people sometimes make it possible for us to give the customer something better for less money. To do this we must improve efficiency by taking advantage of new discoveries in the properties of matter, the sources of power, and the customer's needs. We then relate these elements to one another by shorter, more direct paths. First we must find out what the customer really wants, whether he is an infantry captain knee-deep in a jungle swamp or a scientist analyzing lunar rocks.

Adjusting for utility involves the delicate process of interpreting customer needs and desires in terms of numbers. This requires considerable training in eliciting information from the customer and converting it into the utility function on the graph. The danger here is that some utility functions coincide with certain well-known mathematical curves. This may lead one to assume that other unknown utility functions will fit neatly into mathematical patterns. Many do not. A most useful explanation of utility for our purposes can be found in Chernoff and Moses.

Setting Upper and Lower Bounds

The objective, more often than not, combines several requirements. Each of these requirements offers us a range of choice between lower and upper bounds of demand. The threshold or lower bound is the "least favorable but adequate condition." This level must coincide with the equivalent condition of the other requirements and with the "passing grade" of the rating scale (usually 70). The upper bound is the "best practical condition." It must coincide with the equivalent condition of the other requirements and with the "very good" grade of the rating scale (usually 90). If 90 is "best practical," then 91 is "excessive," and 100 indicates the absolute maximum. These extra benefits can be accepted when they are not achieved at the expense of actual requirements within the working range.

A Word of Warning

As we approach the deceptively pat techniques of mathematical evaluation, we must beware of misreading the relationship among the requirements. One such pitfall is assuming additivity without the proper safeguards.

THE ADDITIVITY ASSUMPTION

Performance, safety, comfort, and style add up to a good car. Antipasto with anchovies, followed by chicken cacciatore, and topped off with spumoni ice cream make up a good Italian meal. The way of a man with a maid, the Bible tells us, is wonderful indeed. The first example is truly additive. The second is additive only in sequence—adding anchovies to the spumoni ice cream may not improve it. The maid is certainly a benefit in "the way of a man with a maid," but two maids may not be better than one.

Recognizing Additivity

Four factors govern additivity among the benefits: compatibility, balance and proportion, sequencing and timing, and interaction. The first two are self-explanatory; the others call for some thought.

SEQUENCING AND TIMING. Though anchovies and spumoni ice cream may be incompatible, proper sequencing can create compatibility. Anchovies as hors d'œuvres and spumoni as dessert, with a main course in between, do add to a good meal. The way of a man with two maids presents no problem if the dates are timed for different evenings.

INTERACTION. Interaction among the benefits is the one aspect of the additivity assumption that calls for most precautions to avoid error.

The three gifts brought by the Wise Men of the East in the Christmas story provide an example of additive values. Let us say the gifts were worth 100 talents each: 100 talents of gold, 100 talents of frankincense, and 100 talents of myrrh. The newborn babe then received 300 talents worth of this world's goods. The figures add up. Now, compare these contributions with the services rendered by the three government contractors in the Arabian fairy tale: the silversmith whose magic mirror revealed that the princess was dying of a wasting illness; the sorcerer, whose magic apple cured the princess; and the rug merchant, whose magic carpet delivered the apple. All three men claimed the award fee.

"Without my apple," the sorcerer argued, "she would not have been cured."

But the silversmith protested, "You did not even know she was sick. If it weren't for my mirror you wouldn't be here!"

"Ha!" shouted the rug merchant. "Without my flying carpet, *neither* of you would be here!"

The happy but bewildered Caliph had to split the reward three ways because it was the combination of the three articles that put the princess back in good health. One hundred talents worth of frankincense, by itself, may still be worth 100 talents, but the magic apple, without the mirror to locate it or the carpet to deliver it, would not have been worth anything to the princess.

The magic articles of our story depend on one another for their performance. When the value of one is reduced to zero, the value of the whole mission disappears, just as the volume of a cube disappears when any one dimension is reduced to zero. Instead of a sum such as $100 + 100 + 0 = 200$, the relationship is more like $100 \times 100 \times 0 = 0$, which of course is a product.

When we understand the exact nature of the relationship among the values we are intermingling, there is no problem. If they are completely dependent and their combined value constitutes a true product, we simply multiply. Say the three magic articles are each 90% efficient: $0.90 \times 0.90 \times 0.90 = 0.729$ or a 73% probability that the princess will get well. In many cases the relationship among requirements or benefits is neither a product nor a sum, though it often approaches a sum, the interdependence being slight or uncertain.

It is possible to bring ponderous mathematical tools to bear on this problem, provided we fully understand the nature and degree of interdependence in each case. A more robust approach is to assume additivity and safeguard the work from serious error by the use of procedures that minimize or exclude the nonadditive portion of mixed values.

Safeguards Against Error

The most important safeguard is provided by the lower bounds of customer acceptance or equipment performance. These bounds have a threshold nature—below them the customer will not buy or the equipment will not work. Introducing into the process of optimization one or more benefits below the threshold level of adequacy or, as happens more frequently, permitting a supposedly unimportant benefit to be driven below its threshold level by the human tendency to maximize the more glamorous ones can be disastrous. We guard against this error by excluding from the COMBINEX scoreboard any input that falls below a previously established level of adequacy.

Another safeguard consists in preventing the introduction of an excessive measure of any one benefit into the COMBINEX scoreboard. Such an excess would displace an equivalent amount of desirable or necessary measures of other benefits. When excessive quality, excessive economy, or excessive reliability masquerade as desirable benefits, they introduce an error that leads to sheer waste.

This error is excluded by the upper bound of the desired benefits. It is not for the manufacturer to tell the customer how much is too much or what he needs or does not need. The customer himself sets these bounds through his buying practices. The manufacturer therefore must learn the upper bounds of customer satisfaction from his own marketing, styling, and quality-control people.

WEIGHTING FACTORS

Assigning measures of relative importance to the requirements or benefits calls for the greatest possible information on the intended use of the equipment and on the goals of the decision maker. It also calls for marketing information on customer wants and resources.

Marketing's Vital Contribution

A value analyst would no more dream of conducting his company's defense in a lawsuit than he would of delivering a baby or celebrating mass; yet he may well sally forth into the world to find out what the customer wants. There are professionals in most companies whose business is just that—finding out what the customer wants, what money he has available, and how to offer the company's products in exchange for some of that money. These marketing professionals command the services of highly specialized researchers on consumer behavior and consumer resources.

In the same way that he calls on other company specialists to help analyze their particular contribution to product value, the value analyst must call on the marketing specialists to help determine the relative importance of customer requirements.

The total importance of the requirements or benefits is set equal to unity, which is then divided by the number of requirements under the initial assumption that they are all equally important. This is seldom the case, but it is a good way to start. If there are five requirements, each weighted at 0.2, some will have to be raised and others lowered.

These measures of relative importance or weighting factors are placed under their corresponding requirements (Figure 8-3). The decision is really made when this is settled, but there is much backing and filling before it is settled. Once the relative importance of the objectives has been determined, all that remains is to find out which of the available choices best satisfied that decision.

EFFECTIVENESS OF THE VARIOUS CHOICES

Each choice or course of action contributes to each of the requirements or desired benefits in varying degrees of effectiveness; for example,

Benefits	TOTAL WEIGHT	HEAVIEST CASE	LARGEST CASE	FEWER UNITS	SET-UP TIME	
Weights	0.30	0.20	0.15	0.10	0.25	
Old design	70	70	70	80	75	Merit
	21	14	10	8	19	72
High performance design	73	80	74	82	75	
	22	16	11	8	19	76
High endurance design	80	74	70	74	71	
	24	15	10	7	18	74
High mobility design	90	89	89	80	90	
	27	18	13	8	22	88

Figure 8-3 Example of a submatrix: elements of mobility. The last column which shows the relative merit of each design's mobility is entered in the MOBILITY column of the COMBINEX scoreboard (Figure 8-4).

system weight and *weight of largest case* make a negative contribution to *mobility*. We want a simple "additive" matrix with nonnegative numbers, so we must convert this contribution into its complement of "lightness" or *positive* contribution to *mobility*, as shown in Figures 8-1 and 8-2.

USE OF SUBMATRICES

The various elements of mobility are combined in a submatrix (Figure 8-3) and weighted for relative importance. The ratings introduced from the graphs in Figures 8-1 and 8-2 are entered in the body of this submatrix, as are the other numbers arrived at in the same manner. These numbers are multiplied by the weighting factor at the top of their respective columns, and the results are entered in the lower right-hand corner of each cell and then added across to yield the right-hand column of "relative merit" in mobility. This column represents the effectiveness with which each choice contributes to *mobility* in the final matrix. Note that the numbers now appear in the upper left-hand corner of the cells in Figure 8-4.

RACKING UP THE RESULTS

The ratings in the final matrix (Figure 8-4) are multiplied by their weighting factors, as before, and then added across to yield the relative merit of each choice.

Suppose that instead of selecting the best of four alternatives we were allowed to make changes resulting in different and perhaps better combinations. As a Persian mathematician once put it, "Could we but shatter this scheme of things entire, and make it nearer to the heart's desire!"

Sometimes we can! Not as a committee investigating someone else's work but as the personal staff of whoever is responsible for the product.

If we provide the man responsible with specialists in the relevant disciplines, other than his own, he can make new decisions involving conflicting requirements and unsuspected rates of exchange, thus appraising the leverage exercised by each element on the value of the whole system.

It is important to bear in mind that scientific aids to decision only align the information for the exercise of judgment by the responsible risk taker. These aids provide neither wisdom nor insight but they do reveal previously unperceived magnitudes and relations which bring into play the wisdom and insight of the responsible managers. For this

Benefits	PERFORMANCE		ENDURANCE		MOBILITY		INITIAL ECONOMY		FIELD ECONOMY		Merit
Weights	0.20		0.18		0.22		0.23		0.17		
Old design	70	14	73	13	72	16	70	16	75	13	72
High performance design	89	18	79	14	76	17	88	20	81	14	83
High endurance design	76	15	89	16	74	16	71	16	87	15	78
High mobility design	70	14	72	13	88	19	75	17	70	12	75

Figure 8-4 The COMBINEX scoreboard as used in selecting a field radar.

reason the COMBINEX, and similar methods, are as useful in *presenting* recommendations as they are in arriving at them.

WHEN THE BENEFITS ARE NEGATIVE

"Less noise!" calls for a negative benefit; its positive form is "A little quiet, please!" Each undesirable feature has an opposite counterpart. The English-language insult—it is not an insult in other languages— which recommends descent into the hottest part of the netherworld has its positive counterpart in the Irish toast, "May you be in heaven half an hour before the devil finds out you are dead!" Instead of talking about a half-empty jug of wine, we can talk about a half-full jug of wine. A recalcitrant maiden, unmoved by verses, bread, or wine, can also be described as a well-fed, virtuous, and challenging girl. *Reduced weight* can be put in terms of *lightness* and *low cost* in terms of *economy*.

The change from negative into positive attributes offers all the advantages of positive numbers; it is easier to add them up and to program the computation.

Identifying the Expected Benefits

Let us say we are ordering a complex waveguide structure that requires a smooth interior surface finish between the bounds of 8 and 64 microinches, where 8 is the smoothest we can use practically, and 64 is the roughest we can accept. Dimensional precision is another critical requirement which must fall between an upper bound of ± 0.0005 inch and a lower bound of ± 0.003 inch. Weight is a penalty because the equipment must be transported rapidly over rough terrain. Cost can be decisive to the contractor because the contract will be awarded on competitive bids and to the user because cost determines the number of equipments he will receive for the funds available.

Upper and lower bounds have been established for economy of production and lightness of weight, the lightest weight practical being 150 pounds and the heaviest acceptable 190 pounds. Initial economy must fall between the bounds of lowest cost practical, $1500, and highest cost acceptable, $5500.

Rating the Benefits

Having identified the benefits and defined their upper and lower bounds, we have a range of choice that allows the free play necessary to arrive at the best combination.

Varying the proportions, after all, is what artists do when they mix paints, what poets do when they combine words, and what cooks do when they combine the effects of moisture, temperature, time, and seasoning on the flavor and texture of well-prepared food.

In the case of our waveguide structure we will have to combine micro-inches of surface finish with tolerances in decimals of an inch, pounds of weight, and dollars of cost.

All of these peaches and pears, apples and oranges contribute to the function of the equipment and we can measure *that*. This example was chosen because the raw measures are negative—something the reader should learn how to handle. Microinches of surface finish are a measure of roughness; the lower the figure, the smoother the finish. Tolerances, being a measure of the allowable departure from a given dimension, are another inverse measure; a lower figure means a more precise fit. The less surface roughness, the smaller the tolerances, the lighter the weight, and the lower the cost the better. These contributions to the function of the equipment are what we want to measure. Two examples are enough to show how the transformation is accomplished:

(a) *smoothness of surface finish,* transformed from the raw data which was measured in microinches of surface roughness (Figure 8-5) and

(b) *dimensional precision,* transformed from tolerances measured in decimals of an inch (Figure 8-6).

The straight line sloping down from the left in Figure 8-5 both inverts the relationship so that low numbers get a high rating and transforms the raw data into the 70-to-90 rating scale.

The curved line also slopes down to the left in Figure 8-6, inverts (downward slope), and transforms (from one scale into another), but, in the course of transformation, it takes into account the difference between tolerances allowed and the *effect* of dimensional precision on the function of the equipment.

ASSIGNING WEIGHTS

There is a balance and proportion among the benefits that will yield the best performance in a product and the greatest effectiveness in a system. One way of describing such a balance, when we know what it should be, is to assign measures of relative importance, or weighting factors, to the benefits as outlined before. Even when we *do not* know what the balance should be preliminary weighting factors serve as a basis for experimenta-tion, calculation, or simulation in the search for the best balance.

Experience with the performance of electronic waveguides which form

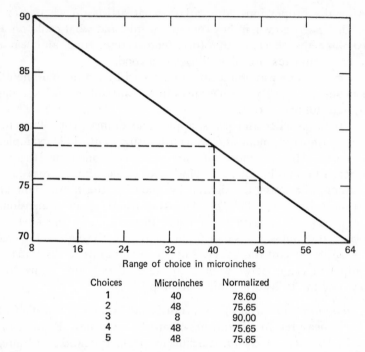

Figure 8-5 Microinches of surface roughness into smoothness of surface finish. The column of normalized ratings is entered under B_1 in the COMBINEX scoreboard.

part of transportable military equipment reveals that dimensional precision is by far the most important consideration, being as important as the other three benefits put together. It is therefore assigned half the total weight, or 0.5 (see Figure 8-7). Smoothness of surface finish and economy of production are each considered twice as important as lightness of weight; so the remaining 0.5 weights are distributed on a 0.2, 0.2, 0.1 ratio, the total weights adding up to unity.

Setting up the Scoreboard

Figure 8-7 shows the COMBINEX scoreboard with all the work completed. The benefits, B_1 through B_4, are listed along the top, with each of the weights beneath its corresponding benefit, and the choices or options, C_1 through C_5, are listed in the column at the left.

Now look at Figure 8-5. Choice 1 has a surface roughness of 40 microinches. The dotted line rising from the 40 along the horizontal axis shows how this number is transformed to its rating of 78.60 on the vertical axis. Choices 2, 4, and 5 all have a smooth-finish rating of 75.65, and

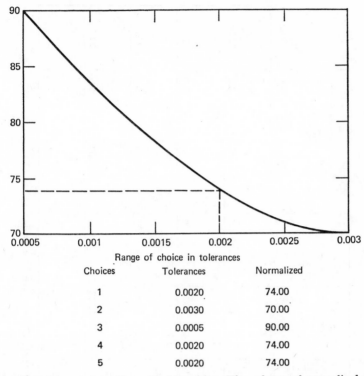

Choices	Tolerances	Normalized
1	0.0020	74.00
2	0.0030	70.00
3	0.0005	90.00
4	0.0020	74.00
5	0.0020	74.00

Figure 8-6 Tolerances into dimensional precision. The column of normalized ratings is entered under B_2 in the COMBINEX scoreboard.

Choice 3 (8 microinches), the best finish, is rated at 90.00. The ratings for the five choices then appear in the column under Smooth Finish in Figure 8-7. They are written in the upper left-hand corner of each box.

The same procedure is followed for the other benefits whose ratings are written in the upper left-hand corner of the boxes under them.

These ratings are then multiplied by the weights above them and the results, entered in the *lower right-hand* corner of each square, are added across to give a figure of relative value for each choice.

Looking at the column headed Relative Value in Figure 8-7, we note that Choice 3, electroforming, rates highest *for this particular application.*

AN INDUSTRIAL ENGINEERING APPLICATION

The richest return in value analysis comes from working on the product itself, leaving it up to the manufacturing professionals to find better ways

Benefits Weightings Choices	Smooth finish B_1 0.2		Production economy B_2 0.2		Dimensional precision B_3 0.5		Lightweight B_4 0.1		Relative value
C_1 STANDARD WAVEGUIDE COMPONENTS SPECIAL JOINTS	78.60	15.72	70.00	14.00	74.00	37.00	72.70	7.27	73.99
C_2 FORMED AND PUNCHED ALUMINUM SHEET DIP OR OVEN BRAZED	75.65	15.13	90.00	18.00	70.00	35.00	90.00	9.00	77.13
C_3 ELECTROFORMED ON FIXTURED MANDRELS	90.00	18.00	78.80	15.76	90.00	45.00	88.80	8.88	87.64
C_4 FORMED AND PUNCHED ALUMINUM SHEET, ELECTRON BEAM WELDED	75.65	15.13	81.20	16.24	74.00	37.00	90.00	9.00	77.37
C_5 INVESTMENT AND DIE-CAST COMPONENTS BRAZED TOGETHER	75.65	15.13	71.90	14.38	74.00	37.00	70.00	7.00	73.51

Figure 8-7 The COMBINEX scoreboard: selecting a manufacturing method.

of making it. Sometimes, however, a value task group provides the best approach to a factory problem, just as methods improvement or work simplification sometimes provide the best way to improve a product and, very often, a service.

The example I cite now started as the value analysis of a high-quality product for industrial use, a product that I will mask under the cryptic designation AMT-4.

How to Improve the Best

A task group had looked at the product from the end-user's standpoint and concluded that quality was what sold this grade of supplies.

"O.K.," said the plant manager, "let's exploit that advantage; let's come up with even better quality—what are you looking at your fingernails for?" he suddenly asked the quality man. "This isn't a manicure session!"

Unperturbed, the quality engineer said, "A gesture of modesty, sir. How can you improve the best?"

"All right, all right," the boss agreed, "but I still want to press our advantages. Let's advance! Let's keep *the others* on the defensive." Enthusiasm made his dark eyes sparkle. "What other advantages has our product got?"

"It's a little cleaner—"

"And it looks good—"

"—and sometimes we sell more than we can make."

Beaming, the plant manager told them, "You got the measure. When a package of our AMT-4 goes to heaven, they'll have to clean up the place to receive it! Let's all go to lunch together and talk about a task group to work on it."

Figures 8-8 and 8-9 describe the team and the task. Figure 8-10 lists the options developed. Guidelines of the selection process are explained in Figure 8-11. Figures 8-12 and 8-13 show how the COMBINEX method was used to select three of the four feasible options.

PUTTING A WHOLE PLANT THROUGH VALUE ANALYSIS

This is the kind of operation most analysts dream about. When it happens, you just can't believe it. At any rate, I didn't.

"Carlos, *mi amigo*," the voice said over the phone, "can you spend two or three weeks in our plant?"

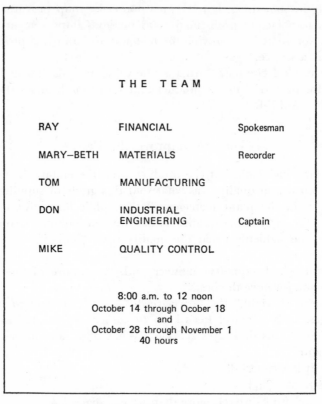

THE TEAM

RAY	FINANCIAL	Spokesman
MARY–BETH	MATERIALS	Recorder
TOM	MANUFACTURING	
DON	INDUSTRIAL ENGINEERING	Captain
MIKE	QUALITY CONTROL	

8:00 a.m. to 12 noon
October 14 through Ocober 18
and
October 28 through November 1
40 hours

Figure 8-8 The task group in a small plant.

"Two or three weeks! . . . *?* Do you realize how many divisions we have? I'll send you—"

"No, Carlos, he wants the Old Tiger himself."

I was intrigued. "Who is this *he!"*

"Our accounting manager. If you are an Old Tiger, he's a Young Lion."

The phone receiver made a whistling sound. Nothing wrong with it; my *amigo* had simply whistled to express his feelings. He was the materials manager in a long-established plant which had been acquired when the corporation took its present form. My *amigo* was a world traveler, accomplished linguist, and as it turned out, adept at bribery.

He was whistling about his plant's manager of accounting. "The guy's built a fire under everybody. He's got us all hopping; and, frankly, we like it."

THE TASK

ANALYZE AUTOMATION OF AMT-4

EXPECTED   BENEFITS

	Weights
IMPROVE CLEANLINESS	0.2
IMPROVE TURNOVER	0.2
REDUCE HANDLING	0.1
ECONOMY OF IMPLEMENTATION	0.5
	1.0

Note: The sum of the weighting factors is set equal to unity.

Figure 8-9 An IE value analysis project.

"Well, I can't go," I told him. "It's just impossible."

"Impossible? Did you get the San Miguel beer?"

"*What* San Miguel beer? I didn't know you could get it in this country." I was making conversation while my girl brought me the folder on his plant. She walked in now, with the folder and a six-pack of San Miguel beer.

"Somebody left this for you," she explained, putting the six-pack on my desk. "It was labeled chemicals."

Nudging me with her elbow, she pointed to a penciled note on the folder. *The grapevine,* said the note, *has it that this plant will be closed down within a year.*

"Labeled chemicals?" the voice on the phone was saying. "That's the

```
FEASIBLE    COURSES

OF    ACTION

A.  INSPECT AND BOX AT SL STATION

B.  AUTOMATIC ENCASING

C.  AUTOMATIC SEALING

D.  AUTOMATIC SORTING
```

Figure 8-10 Options (not mutually exclusive).

```
            Rating the contribution
          made by each option toward
        the attainment of each objective


YARDSTICK:            A bounded interval scale
                      ranging from 70 through
                      90.

UPPER BOUND (90):     Represents the most
                      favorable yet practical
                      contribution.

LOWER BOUND (70):     Represents the least
                      favorable though still
                      satisfactory contribution.



Note.  By excluding the excessive (over 90)
and the inadequate (below 70) the bounded
interval scale justifies the assumption that the
effectiveness of the contributions is additive.
```

Figure 8-11 The COMBINEX Aid to Decision: ground rules.

Benefits Weights Options	Improve cleanliness	Improve turnover	Reduce handling	Economy of the change	Relative value
	0.2	0.2	0.1	0.5	
Inspect and box at SL station A	90 18.0	82 16.4	88 8.8	88 44.0	A 87.2
Automatic encasing B	90 18.0	88 17.6	86 8.6	73 36.5	B 80.7
Automatic sealing C	82 16.4	78 15.6	84 8.4	90 45.0	C 85.4
Automatic sorting D	90 18.0	80 16.0	86 8.6	70 35.0	D 77.6

Figure 8-12 Evaluation on the COMBINEX scoreboard.

SELECTION

Options *C* and *A* look substantially better than *B* and *D*.

D is set aside for future implementation.

B is recommended along with *C* and *A* because *C* and *A* depend on it.

Figure 8-13 Recommendations.

six-pack I sent you. The rest of the case is waiting for you when you come
to see us."

How to Handle the Impossible

The San Miguel beer was followed by more chemicals: Carlos Primero
brandy, Marques de Riscal Rioja red, Spanish anchovies, and finally, the
decisive blow—*chorizos!*

Chorizos are essential for authentic Spanish rice. Imagine pepperoni,
smaller in diameter and succulently tender. Of course, I went—every
man has his price.

After a four-hour drive I arrived at the motel ready to take a shower
and call for room service. In the lobby three men converged on me, the
materials manager, whom I knew, and what I took to be two gigantic
bodyguards. But they weren't.

One of them turned out to be the Young Lion, and the other was
the systems and procedures analyst. The Lion took over and became the
man in charge. It seemed natural.

"We got you a good room," he said, "follow me."

We all followed, and a waiter followed us, pushing a cart loaded with
food and drink.

In the room the manager of accounting smiled at the group the way
Napoleon smiled at the surprised generals when he took over the Army
of Italy.

"Now we have the full team," he announced, pouring me a drink.
"You'll be proud to quarterback it."

If I was the quarterback, I thought, this man was both coach and
owner. With the help of the plant manager and the manager of warehousing and distribution he had already organized and launched a massive and up-to-the-minute value-analysis program.

"Now that the program is rolling," the man in charge continued, "we
want the real professional to help us apply dynamic value analysis to all
segments of operations—such as systems and procedures, capital equipment, material and labor, and—"

I was already beginning to feel both dynamic and necessary. That is
how Napoleon had won acceptance from the Army of Italy. He made
everybody feel important, and important they became.

"Because it will take a professional like yourself," the man in charge
continued, "with all the prestige of corporate headquarters behind him
to show our people how much they can really do."

While the materials manager turned down the bed covers, the systems
and procedures analyst unpacked my bag.

"We will let you rest now," said the man in charge, setting my alarm clock for five-thirty in the morning. "You are scheduled to meet with the plant manager, the manager of warehousing and distribution, and their staffs *at seven*."

The systems and procedures analyst said, "We'll meet you in the lobby about five of."

I had to remake the motel bed to take care of my increased professional stature. If Henri Fayol had descended from heaven, with an angel on each side, he could not have been treated with greater courtesy and deference.

At the meeting next morning the manager of accounting took charge with the ease and aplomb of an admiral's flag captain. "You all know Carlos, from corporate staff"—he made *corporate staff* sound like The Vatican—"Carlos is going to apply his COMBINEX Aid to Decision to our master plan for operations improvement. That way we can concentrate the effort where it will do the most good. Now, let's bring him abreast of the situation."

Everybody explained the situation.

"They've built this huge automated plant in the Midwest—"

"And, of course, it costs them less to make the product—"

"And to warehouse it—they are up to their ears in conveyors and automatic controls."

"Our machines can't keep up with theirs. Our machines are 50 years old!"

"And our girls are 50 years old!"

"And the home office wants to shut us down, but none of us want to move to the Midwest—"

"We got grandchildren here. Why do they want to shut us down when we are meeting our standards?"

"Ha!" interrupted the manager of accounting. "Tell Carlos about the standards. *What* standards are we meeting?"

Someone brought me a cup of coffee.

Putting it down, he spread his arms in helplessness. "The best that 50-year-old machines and 50-year-old girls can do. That's our standards."

"Yeah, we're doing the best we can and they want to close us down."

"And we don't want to move to the new plant. We shouldn't have to if we are meeting standards."

The accounting manager spoke quietly, and everybody stopped to listen. "As of the first of the month," he announced, "our standards are the same as those of the new, automated plant—"

Hubbub ensued: "Impossible! Can't do it! They're automated; we ain't automated! Not fair! It'll put us in the red!"

"If we are not competing with another plant," continued the accounting manager, "we *are* in the red. As for 'impossible' and 'can't do it,' that's beside the point. The situation boils down to Kant's categorical imperative—"

"Kant's *who?*"

"Who's Kant?"

"Guy back in Germany," explained the accounting manager. "Kant wrote, 'I can because I must!' Boys, it is not a matter of we can or we can't, *we've got to!* To stay in business *we've got to be competitive,* agreed?"

The bosses nodded agreement, and someone said, "I guess so, them Germans, like Kant, are pretty good production people."

A Value Task Group at the Top of the House

The bosses personally plunged into the program. "If you intend to train middle management and supervision without training us, you are going to have trouble implementing your ideas," warned the plant manager.

"That's right," agreed the manager of warehousing and distribution, "we ought to know as much about value analysis as the rest of you; otherwise how can you have an effective program for the whole location."

The outcome was that the two top bosses attended all the lectures, participated in all the planning sessions, and became an integral part of the value-control group, together with the accounting manager, materials manager, and systems and procedures analyst.

The task of this group was to plan, organize, and direct a broad-front value effort which was deployed in eight teams, each made up of five supervisors selected from five different departments to provide a balanced combination of skills.

The Function of Manufacturing

The function of manufacturing is often described as *produce goods* and *make profit,* but this description leaves a great big gap, as the value-control group soon discovered. Before manufactured goods can be turned into profit, *somebody has to buy them.* When this point came up, I had the pleasure of watching the control group in action.

"The trouble is," said the plant manager, "that we have little to do with sales. We are in a quiet, out-of-the-way area, with a nice labor pool and convenient freight transportation, but the marketing people are all in the big cities. What can we do, by ourselves, to make the customer want the product?"

The materials manager grinned, his eyes twinkling. "Why don't you ask me?" he suggested.

"You've worked in Marketing?"

"No, but I've spent most of my life buying. Buying is what customers do, isn't it? I have a buying problem right now. Did you see that little note that was brought to me a while ago?"

He took out the piece of paper and read it: "No sign of the truckloads of raw materials. If they don't get here by noon, we'll have to shut down six presses."

The plant manager was on his feet. "Why didn't I—" he thundered incoherently. "Nobody . . . *YOU* should've told me—where's that foreman!"

The materials manager raised a restraining hand. "Cool it, sir. We are on top of it—"

"I don't want people on top of it; I want the trucks!"

"Look out the window," suggested the materials manager.

"Oh . . . but why are they late?"

"New drivers. Tried to deliver the stuff at the Lutheran Church. Minister shooed them down here."

The plant manager looked hurt. "Why did you get me all upset?"

"I sure didn't mean to get you upset," said the materials manager, "but it was the best way to answer your question."

"What question?"

"About what we can do, by ourselves, to make the customer want the product."

"What does that have to do—"

The materials manager smiled broadly. "You were acting like a customer a minute ago, and you sure thought *delivery* was important."

Opportune Delivery

Opportune delivery entered the picture as a contribution to the worth of the product. As the warehousing and distribution manager put it, "A good product is not much good if it is not there. Missing delivery dates means losing customers." He was the one who insisted on the word *opportune*. "Opportune means *when* you want it, not before or after. Making a product before it can be sold or used means that somebody's money is tied up in useless stuff that occupies costly warehouse space and deteriorates in storage *and* could go out of date."

Credibility

Credibility emerged as a consideration when the systems and procedures analyst pointed out that *meeting dates* meant *keeping promises* and that there are two aspects to a promise, *making it* and *keeping it*. "Boy, that's one for *me* to work on," he added. "Here is a contribution to the list of projects—better communication between the makers and the keepers of promises."

Accounting and systems and procedures had already conducted a joint study of the credibility factor, of the relationship between *promise* and *delivery*. They identified elements of this plant's promises as WHAT, HOW MANY, HOW GOOD, WHEN, and FOR HOW MUCH.

The WHEN was discussed under opportune delivery; it refers to elapsed time. WHAT and HOW MUCH referred to making the right quantities of a given product for a given customer. Both WHAT and HOW MANY were put under the heading of . . .

The Right Quantity

The value-control group agreed that the right quantity for the customer is the usable quantity he receives, and not the quantity shipped. "There is a lot of time lost," pointed out the manager of warehousing and distribution, "between the discovery of goods damaged in shipment and their replacement. The blame and the payment for the damage are small matters compared with what we and the customer suffer when our goods are not available as ordered."

Someone suggested a statistical analysis of goods replaced after shipment. The suggestion was added to the list of projects.

The right quality refers to the element, *how good.* The whole control group agreed that maintaining and improving quality was a major consideration. "Quality is like a happy marriage," the plant manager said, "you have to work at it or it spoils."

The Right Price

Everybody agreed that price was determined by the customer and the competition. The pricing office simply determines whether we can afford to make the product at that price.

"It seems to me," said the materials manager, "that you are all looking at our product the way I look at everything I buy: opportune delivery, the right quantity, the right quality, at the right price. I also take a pretty good look at the service I get from the supplier."

After discussing the concept of service the group agreed that the word *service* really encompassed everything contained under the "sales promise": delivery, quantity, quality, price, and standing behind the product.

Using the Combinex Scoreboard

The systems and procedures analyst had gone over to the blackboard and was drawing the framework that appears in Figure 8-14. Now he printed the word SERVICE at the head of the first column.

TIMELINESS. Timeliness in the sense of opportunity, such as timing the injection into orbit of a space vehicle, was discussed as thoroughly as *service*.

FEASIBILITY OF IMPLEMENTATION. This included all the implementation requirements discussed at the end of Chapter 9.

SAVINGS. Savings were considered as important as pumping bilges on an old ship. Until the payload yields enough profit to pay for drydocking, pumping bilges must be a major consideration.

Benefits Weights / Projects	Service	Timeliness	Implementation	Economy	Relative value
	0.35	0.30	0.20	0.15	
001 On line real time communication	90 / 31.50	90 / 27.00	90 / 18.00	84 / 12.60	89.10
002 Improve test and measurement equipment	90 / 31.50	90 / 27.00	85 / 17.00	80 / 12.00	87.50
184 Relocate ovens	60 / 21.00	60 / 18.00	60 / 12.00	60 / 9.00	60.00
185 Improve visibility in passageways	60 / 21.00	60 / 18.00	60 / 12.00	60 / 9.00	60.00

Figure 8-14 A COMBINEX comparison of 185 projects.

WORKSHOP. The workshop was really a major offensive against adversity: six weeks of concentrated effort Mondays, Wednesdays, and Fridays from 4 to 6 P.M. and six months of follow-up, meeting twice a month.

Each of the eight task groups followed its own job plan. After all the creative sessions were completed 46 participating supervisors had generated 200 workable ideas, of which 185 were put through the COMBINEX method. Figure 8-14 shows the first two and the last two of the 185 options after the computer had ranked them.

Results

After the first 18 options had been implemented the plant was out of the red. The following year people in the Midwest began complaining that automating the old plant was not in the five-year plan.

This was too much! I had to go and see what they were doing. The plant was humming all right, but it was the same old plant.

"You are meeting dates," I told the materials manager, "you are getting more business, and you cut production costs."

He beamed.

"But," I continued, "these are the same 50-year-old machines and the same 50-year-old girls. What *did* you do?"

"Oh, shucks, it was nothing, Carlos. We just rebushed the machines and we let the girls *think!* There's a powerful lot of brains in a factory full of women."

APPRAISING VALUE-ANALYSIS OPTIONS

Among the options developed in a value-analysis study one or two may stand out as superior to the others. In that case simple inspection of performance and cost figures reveals the best one.

The COMBINEX Aid to Decision should be used only when aid is needed, when choice is difficult, when there are a great number of choices —as in the example just cited—or when the reasons for the choice must be explained to people who have not participated in the study.

Up to this point we have been concerned with providing a better product for the money or providing the same product for less money. In any case, we have been studying the product itself or, more accurately, what it does for the customer.

Now we have to provide information on the merits of our proposals themselves.

Benefits from Whose Point of View?

Before value-analysis proposals can do anything for the customer they must do something for the company; and before they do anything for the company they must do something for the people who approve them and implement them—the risk takers.

Who must approve our proposals? What are his needs? What are his problems?

Who will allocate the funds? From what source?

Who will do the work? When? Where?

What department will gain the most? Are they represented in the task group? Do they understand the benefits they will derive?

What department will suffer? Disruption? Loss of workload? Overload? Schedule delay? Loss of control?

Answering these questions, and many more like them, will reveal advantages and disadvantages associated with a given proposal. Earlier comparisons helped us to select options that benefit the customer. The present comparison will help us to select from among those options the ones most advantageous to our management.

Advantages and Disadvantages

In appraising advantages and disadvantages, the first pitfall to avoid is counting them. Their *number* is not what matters. One single advantage such as being a pretty girl often outweighs many disadvantages in a secretary. "Oh well, you can always teach them to read and write," is often given as justification for the obvious choice.

A single disadvantage, too, can outweigh all advantages, such as the speed, endurance, and spirit of the horse Borysthenes, who has the disadvantage of being dead 1900 years.

The second pitfall to avoid is canceling out disadvantages against equally important but unrelated advantages, as was done by the merchant of Ahmadabad who claimed full price for a slave girl, saying, "She is missing two fingers, but she has 12 toes."

The mechanics of the COMBINEX method seldom need to be applied to the subsequent advantages and disadvantages of accepting a proposal because COMBINEX ground rules have already established certain safeguards, such as setting bounds. In the foregoing examples judiciously set bounds would have excluded illiterate secretaries and dead horses.

Disadvantages cannot be inverted too meaningfully into positive bene-fits. By definition, disadvantages are incorrigible baddies. The disad-vantage of schedule delay would have to be inverted into *less schedule delay*. The phrase has a hollow ring. Risk would have to become *safety, security,* or, would you believe, *invulnerability?* Yet we had no trouble inverting surface roughness into *smoothness of surface finish* or tolerances into *dimensional precision*. Why the difference?

Two Different Kinds of Animal. Design or service requirements must have both an upper and a lower bound, but once a satisfactory mix has been achieved its *advantages* are usually open-ended at the top, provided the improvement does not add to the cost. If being pretty is an advantage in a secretary, we do not mind if she is *very* pretty or quite beautiful. If dollar gains are an advantage, we do not mind how much *more* money we make.

We must set *lower* bounds, however, at the point in which an "ad-vantage" ceases to be an advantage. If the appearance and manner of secretarial applicants ranges from "she is delightful and divinely beauti-ful" (decided advantages) down to "she is mean and her face stops clocks," the lower bound should be set no lower than "her appearance and manner cheers you on Monday mornings."

The concepts of utility and effectiveness keep reappearing in this book when there is a difference between an input, such as the optical image of the girl, and the effect of that input. Let us say that there is a clock-stopper who has a very pleasant manner, likes her work, and is happy to see her boss. He could well rate her as a girl who "creates sun-shine on a rainy Monday morning," cheerfully adding, "Who needs clocks!"

If advantages are open-ended at the top, disadvantages are open-ended at the bottom. We do not care how small they get, and we are happy if they disappear altogether, provided reducing them does not add to the cost. At the top, however, they cannot be open-ended. We do not want disadvantages *growing* beyond unacceptable limits. We may rate risk simply as *low, moderate,* and *high,* but then we must set an upper bound before *unwarrantable;* say at *very high.*

IMPORTANCE OF TIME, MONEY, AND RISK

At this stage we have a list of advantages and disadvantages, all open-ended at the favorable end and all bounded by the least favorable condi-tion acceptable. We know that some are more important than others and that we cannot match disadvantages against unrelated advantages.

Matching Related Advantages and Disadvantages

This is another story. We *can* match them when they differ only in polarity or direction. Dollar losses can be matched with dollar gains *within a given time frame*. Time lost can be matched with time gained within the same cycle, provided the change does not alter necessary sequence.

An Economic Definition of Risk

This is the probability of an unfavorable event multiplied by the cost of that event. An even chance of losing $1000 yields an expected loss of $500.

Suppose we estimate $1000 as the cost of implementing a proposal, provided we do not run into problems. If we do run into problems, and there is an even chance that we will, it may cost twice as much.

Running into problems would raise the cost of implementation $1000 to $2000, that is, $1000 more than anticipated. Chances of that happening are 50%, so we add $500 of risk capital to the estimated cost of implementation.

The Three Baskets

Most advantages and disadvantages can be put into three baskets—time, money, and risk. The relative importance of these major categories in the eyes of management varies in each situation. In selecting proposals the value task group must ascertain the current importance of each of these three baskets and of the business considerations included in each.

Table 8-1   Day-to-Day Business Considerations

Time	Money	Risk
Cash flow	Sales	Loss of opportunity
Flow of costs	Gross margin	Loss of time
Manufacturing schedule	Reserves	Loss of money
Timely delivery	Profit	Loss of prestige

Value analysis represents a concentration of effort by a particular mix of skills. The mix is determined by the requirements of the project. In analyzing the project, a value task group develops a number of options, one of which may well be to let the product alone because it is already too good to tamper with. Generally, however, the options are opportunities for improvement. Selection of the most fruitful opportunities depends on two distinct factors: (a) the expected value of the opportunity itself and (b) the business consideration most likely to profit from that particular opportunity. These considerations are the compartments of our three baskets, as listed above.

The question then becomes, "Which business consideration is most important now, and which of our proposals will contribute most to that particular consideration?"

Now the reader may well ask, "How in the world can a value-analysis proposal contribute to *reserves!* . . . ?"

1. By reducing service under warranty, through greater product dependability.

2. By modernizing existing equipment, as an annual expense, instead of buying new equipment and depreciating it over the years.

3. By improving capital equipment to extend its useful life.

In addition to saving time and money and reducing the need for reserves, a task group truly concerned with value will identify areas in which additional money invested or additional time spent will yield a return as good or better than routine investment opportunities.

In summary, screen your proposals for their effect on current business needs. Long-term needs were considered at the time of project selection. What can be done to the projects *now* is governed by today's capabilities and requirements. Before preparing recommendations to management, therefore, the value task group should make one final check with Marketing and Finance. This move is part of the strategy of implementation discussed in the next chapter.

IX

IMPLEMENTATION

A good value-analysis proposal constitutes a plan for the concentration of resources on a given opportunity. Such a plan is strategic rather than tactical. Humble as the project may have been, its value has been analyzed from a company rather than a departmental viewpoint. A variety of company resources must now be concentrated on a company opportunity; yet the plan has been developed and is presented by a group of specialists at the working level. It is a strategic plan offered by soldiers drawn from the firing line—not by staff officers but by the equivalent of a squad leader, a gunner, a tank commander, and a helicopter pilot.

Is that bad?

The soldiers of Washington and Bolivar were more effective than their opponents *because* they understood and contributed to the strategy of their commanders. Facing the task of implementation, the squad leader in industry must reach back into his field pack and draw the marshal's baton which, as Napoleon assured us, is surely there. Grasping it and wielding it is a matter of thinking in terms of company strategy.

STRATEGY OF IMPLEMENTATION

Whether it is in war or business, the fundamentals of strategy are the same—practical rules based on experience. The martial classics written

by Xenophon, Sun Tzu, Marshal Saxe, and von Clausewitz were the
work of eminently successful professional soldiers. The elements of
strategy which they identified were learned in a lifetime of fighting and
winning battles. It was a businessman, however, the fabulous Confeder-
ate cavalry commander, General Nathan Bedford Forrest, who put the
basic elements of strategy into one phrase: "Getting there first, fastest,
with the most men." Movement, timing, speed, and concentration.

A value-analysis recommendation which includes movement—in the
sense of change for the better—and concentration of the right skills,
should offer an opportunity for management to make the product
profitably competitive. On the face of it nothing could be more welcome,
but grasping opportunity calls for the following:

DECISION. Which opportunity to grasp? Every morning an operating
manager is faced with a number of opportunities, each supported by
enthusiastic sponsors. He has to decide which opportunities to grasp
and which he must allow to escape.

UNEXPECTED EFFORT. If it is routine, it is not an opportunity; the regular
departments can handle it within their budgets and schedules.

UNEXPECTED EXPENDITURES. If, on a hunting trip, you happen across a
stream full of trout, you may have to buy a fishing rod and bait.

Decision

The first and most important element of the decision process is the
element of value.

The Element of Value. The value system of the decision maker determines whether a decision needs to be made at all and it colors the worth
of the expected benefits. But the whole book is about this kind of value,
so I will go on to discuss the other major elements of the decision process.

The Element of Time. Let us stop for a moment to think about the
nature of time with respect to the sequence of events. In this context
we perceive at once the relation of time to chance. Those events that
have already taken place *before* the instant of observation are certain
and inalterable, although they may be unknown. Those events expected
to take place after the instant of observation are uncertain and may not
happen at all, but events taking place at the very instant of observation
can be controlled. They effect the outcome of future events and they
constitute both opportunity and risk.

The Element of Chance. Although we do not know what the future will bring, we do know that everything we do or fail to do affects the future. We also know what the future might bring, and we know this with different degrees of likelihood.

Like the command of early Roman armies, which was shared by two consuls, command of our destiny is shared with us by chance, which controls the future. But we control the present. In the here and now we face risk, we face opportunity, and we commit resources. Then chance steps in with the consequences. So we must (a) recognize and appraise risk, relating it to the consequences of failure; (b) recognize and appraise opportunity, relating it to the consequences of success; (c) recognize and appraise our resources, commiting them in proportion to the importance of the expected outcome; (d) relate risk, opportunity, and resources to time and space in order to achieve our ends through timely, decisive action.

Students of management science complain that industrial executives, when they let their hair down, admit that their own personal risk is the first thing they consider when making a decision. This is as it should be. Life on earth follows the strategy of survival and growth, and survival comes first. An executive cannot contribute much to his company's growth if he has been fired. A company cannot grow if it has been wiped out. For this reason the strategy outlined above must consider the risk first.

The Element of Conflict. In manipulating the forces of nature man has proved a master. He decides the conflict between soil building and erosion by turning rampaging rivers into placid lakes to provide hydroelectric power. Out of the conflict between fire and water, man gets the steam engine. Even when it comes to conflict with other men he has made minor progress, moving human conflict into courts of law where the lawyers strive to exclude shaky evidence and wring out the truth.

Anyone who has used jumper cables to connect the battery of one car to that of another knows that there is conflict between the positive and the negative poles in an electric circuit. This conflict is called *potential.*

The conflict of a mountain torrent becomes *potential* at a hydroelectric station. In the same way conflict between a husband and wife may appear as *potential* to a divorce lawyer.

In evaluating choices a value task group identifies possible sources of conflict in order to reduce or eliminate unnecessary conflict, utilize the potential in wholesome conflict, and follow the principles of the new discipline of fluidics to convert turbulence into guided force.

Having gone to some length to study the decision process, we go now

to the other two requirements for grasping opportunities, the need to
provide unexpected effort and to incur unexpected expenditures

Unexpected Effort

Unexpected effort means that no one is specifically assigned to do a
task. If it does fall into someone's area of responsibility, he does not
have the time to do it. The plan of implementation, therefore, should
have cleared beforehand: what has to be done, who will do it, and how
much time will be allotted to the task. It should also propose a schedule
for doing the work, for reporting progress on the part of the troops, and
for following up on the part of management.

Unexpected Expenditures

So you buy a fishing rod and some bait in order to take advantage of
the stream full of trout. This illustration is typical of the legitimate
scrutiny given value-analysis recommendations.

You bring home the trout to put in the freezer, and since the family
is going to eat the trout you charge the fishing expense to your wife's
house money.

Her first question could well be, "Did you spend all that money just
to do my shopping for me or because you like fishing?" After figuring
the cost on pencil and paper she would probably add, "I can get the
same fish at the market for half the price, and I don't have to clean
them or freeze them. So have a cook-out with your friends, but not on
my house money!"

A perfectly reasonable supposition on the part of the hard-nosed
guardians of cash is that the proposed improvement may have been
recommended because the task-group members like fishing; that is, they
suggested the change to indulge their perfectionism.

Before engaging in fishing for profit you would check the market
price of fish and your cost of catching them. Before spending much time
on a value-analysis recommendation you must have checked the probable
cost of implementation. All this economic information must be checked
and rechecked to make sure the fish are worth the cost of bait.

HELPING MANAGEMENT CHOOSE WHAT TO IMPLEMENT

Like Colombian coffee implementation is a selective harvest. One of the
advantages of mountain-grown coffee is that only a few berries mature

at one time, thus giving the Colombian coffee grower an opportunity to select only those berries worthy of his famous product. Being best and staying best is often a matter of picking and choosing.

A plan for implementation must provide adequate time and sufficient information for picking and choosing among several opportunities. Mountain-grown coffee does just that. The ripe berry, containing two coffee beans in the center, is the size and color of a maraschino cherry. In a hot climate, unlike that of the Colombian Andes, the entire crop ripens at once and thousands of migratory laborers move through the fields hurriedly harvesting the berries—*all* the berries and anything near them.

Not all tropical climates are hot. What makes them tropical is that the temperature is the same all year around. Eternal snow covers the peaks of the Colombian Andes, and halfway down the climate is an eternal springtime. Here coffee develops at leisure, the same shrub having flowers, green berries, and ripe berries at one time. The farmer and his family have the whole year in which to select only the best berries.

In the manufacturing industry a similar selection has to be made among the many opportunities for investing company time and money. A plant manager has very limited reserves. He must commit them first to plugging those holes that may sink the ship. The remaining reserves can be committed only to the very best ideas for improving the business; and the field includes ideas from many sources other than value analysis. Here, too, we can learn something from the way good coffee is grown.

Providing Enough Information

Mild coffee is not good coffee because all the berries eloquently persuade the farmer to pick them but because they sportingly provide an opportunity for the farmer to select only the best ones.

Nothing is gained when a value-analysis idea is pushed into implementation at the expense of an idea that could contribute more to the company's goals. The value-analysis proposal must be prepared to compete, and to compete honestly, with many ideas for improving the position of the business. Like the mild-coffee berry it must sportingly provide the means for making a good selection.

The first thing that comes to mind when the coffee farmer and his family make their daily selection of ripened fruit is that they have the *time* to do it. Second, they have enough information—information provided by the coffee berries themselves. The vivid red color, the size, shape, and plumpness of each berry can be appraised at once. The farmer does not have to conduct an interrogation, saying, "Berry, are you *really* ripe for picking? Is your chemistry actually complete? Will the beans be of the right size and shape?" Nor should management have to conduct a similar interrogation on a value-improvement proposal.

Making an Honest Forecast of the Consequences

The proposal itself should include sufficient, believable information for an estimate of gains, costs, risks, and payoff period. The estimate itself should be presented early enough for the risk taker to compare the supporting information with the well-documented data that usually accompanies the more traditional proposals.

In addition to the cold numbers of the estimate, the proposal should include an honest appraisal of the effects of implementation on the goals, work schedule, and organizational relations of the man who has to say yes or no.

This man may be your own boss, whom you should have sized up pretty well, or he may be a well-known, easily understood, and quite predictable department manager. Let's say, however, that the decision maker in this case is a sphinx—completely inscrutable—the general manager of your facility. In the past you have communicated with him only in the form of a respectful shudder as you passed him in the halls. He has never heard of you, but out of courtesy to your supervisor he agrees to listen.

After you have blurted out the first sentence and gulped twice, The Boss interrupts you. "Hold on there," he says with upraised hand. "Don't give me any of the details. Take your proposal to Joe Doakes, I want to know what he thinks about it. Tell him to check it out with Dick Roe in accounting."

And you walk out to unload a pile of unscheduled work on Joe Doakes.

Where did you go wrong?

You overestimated the credibility of your information and the value of your judgment in the eyes of The Boss.

The day will come, as it does to all good value analysts, when the top man will say, "Don't give me the details, just tell me what you want to do, and what it takes to do it." That will come only after your information has proved accurate, and your judgment sound, enough times to earn his confidence.

Finding the Right Sponsors

There are other people in the plant who have already earned that confidence—the Joe Doakeses and the Dick Roes. Their support and favorable appraisal are stepping stones to successful implementation. It is necessary to identify the trusted advisers of the decision maker and to bring them on board early—from the very selection of products to be analyzed. Later, when the plan of implementation is being prepared, these men can study it with some foreknowledge of what is intended. They can then suggest changes and modifications to suit business conditions as seen from their points of vantage.

IMPLEMENTATION REQUIREMENTS

To *implement* means to accomplish, to carry out, *to fulfill*, which is exactly what *implere*, the Latin root, means: fill up to the top, complete; in short, get it done. The concept is as old as those Roman roads and Roman bridges still in use today. The Romans were not content with finding the best routes; they built the roads and the bridges, garrisoned them, maintained them, and reaped the benefits for centuries.

Implementation converts planned action into actual action followed through to completion. The report described in the preceding section was, in effect, a plan of action. The other requirements follow.

Authorization

When a missionary says to his Jivaro Indian interpreter, "Tell the chief to make his people eat the food we brought," he is wasting his breath. The "chief" cannot *make* his people do anything. Among primitive peoples, authority is diffused. Warriors choose to follow a leader in the hunt and in war and that is why he is the leader. He cannot, however, tell them what to eat or where to build their huts. Tradition tells them what to eat, and the elders, on the basis of their experience, decide where to locate the village.

The pattern of authority in industry is about halfway between tribal command-and-persuasion and military command-and-control. The industrial pattern, like the tribal pattern, utilizes more of the knowledge and judgment of the entire organization. It can afford to do it because its "commands" are generally more acceptable to the employees than the military command to go out in front of all those bullets.

Project or Recommendation:						
WHAT must be provided or accomplished?	**WHO** will do it?	**WHERE** will it be done?	**HOW** will it be charged	**WHEN** will it be done? Production weeks Write date in line below		
PLANS AND APPROVALS	Authorization						
	Forecast						
	Plan budget						
	Approve budget						
	Plan schedule						
	Approve schedule						
	Plan facilities						
	Approve facilities						
COMPONENTS	Design						
	Drafting						
	Laboratory tests						
	Material requisition						
	Procurement						
TOOLS	Budget						
	Design						
	Procurement						
	Manufacturing						
OPERATIONS	Fabrication						
	Assembly						
	Inspection						
	On stream						

Figure 9-1 Left sheet: implementation.

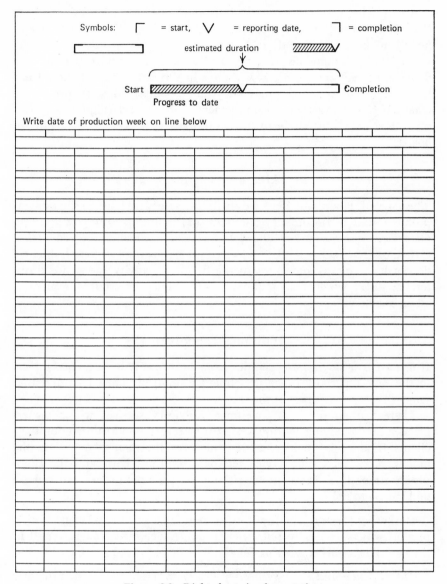

Figure 9-2 Right sheet: implementation.

A surprisingly large number of industrial employees, however, believe that all they need to do to get something done is convince the chief and he will make the Indian do it. You do have to convince the chief or he won't let the Indians do it, but you also have to convince the Indians. Since you have some of the key Indians in the task group, it should not be too hard to convince other key Indians and then work up to the chief. A Gantt chart of the implementation process (Figure 9–1) can serve as a check list to identify the Indians that must be brought on board.

Skills and Facilities

The cost of labor and materials will be considered as an investment problem under *Additional Resources*. Skills and facilities *as limiting factors* are another story. The dollar return from an improved hydrocarbon-oxygen fuel cell, right now, could be ever so attractive, but we do not have an electrochemist in the house. A gravity system for transferring slurry would save pump wear and downtime, but our plant is all on one floor.

Additional Resources

If certain cost savings are pennies from heaven, all investments are pie in the sky—especially if the investment is someone else's idea and that someone else expects part of the credit.

There are hundreds of investment advisory services whose owners make a modest living by discovering investment opportunities for their clients. Those who make the big profits are the men who put up the money. The investment service gets paid for the advice, and the client profits from the investment by risking his own money. Where can we find money to risk?

Marketing Money. The funds value analysis needs to make an existing product sell better, or to put a new product idea on stream, are still in the customer's pocket, waiting for the new or improved product.

The Chicken and the Egg Money Problem. The investment promoter says, in effect, "Give me money to hatch this egg and I will give you a very fine chicken." A value-analysis recommendation, after all, is only an egg. Now who will risk his money to help us hatch it?

The general manager?

Logically, in a sense it is his money, but he is responsible for much more money, most of which he spends through well-proved, usually profitable channels, leaving the specifics of smaller, internal investments to specialists in each area. So, if we want money to hatch chickens, we have to find out who is in charge of chickens.

Our recommendation should yield money from increased sales. What department is responsible for *that*? . . . Marketing of course. They have money expressly earmarked for generating more sales. So you say, "Here is an exceptionally promising egg. Give me some of your incubating funds and I will give you a very fine chicken."

The answer may well be negative: "If you are going to invade our area of responsibility by laying your own eggs, you can damn well hatch 'em yourself. *We* are supposed to generate more sales. We will use *our* money to implement *our* ideas. Tell your mother we don't need any little helpers."

Could it be that we used the wrong approach?

Yes; from way back!

Following the principle of responsibility, the moment the task group discovers an opportunity to improve sales, it should bring Marketing into the act. The Man Responsible in Marketing can then avail himself of the multidiscipline group to generate marketing ideas. If he—as the risk taker—thinks the ideas have merit, he will implement them out of his sales-improvement budget.

If you want to hatch your eggs in his incubator, you have to lay them in his barn, and when the chicks hatch *his* rooster should crow.

Direct Labor Funds. The payroll account is one of the two major valves governing the outflow of cash. The money is budgeted, funneled from company income into pay envelopes, and goes out the door on pay day. A product change that reduces labor cost could keep some cash in the house by eliminating one or more pay envelopes, by "taking an operator off the line."

Such savings are perfectly tangible, will stand audit, and make impressive reports. Sometimes they are even good for the company, but "taking an operator off the line" must be considered as a trade. One pay envelope stays in the house and one trained worker goes out the door. Add severance pay and lowered morale to the exchange and the company may not come out ahead.

Eliminating additional direct labor, however, is fairly clean-cut. When the payroll budget includes people to be hired or reassigned for a particular project, a reduction in direct labor simply means not hiring or transferring the unnecessary operators—tangible savings at little cost in turmoil and efficiency.

Here is an example. A company gets a contract for making a large number of complex electronic systems of its own design. The value task group determines, from the pilot run, that more time and money will have to be spent on "systems test" than in making and testing all the components of the system.

"What does *systems test* mean on this equipment," asks the cost estimator.

"That the factory has to make it work! That's what it means," snaps the factory test engineer.

The buyer in this task group happens to be one of those pleasant and beautiful girls with which Providence sometimes blesses good Purchasing Departments. "But I thought that—" she starts to ask the electrical engineer.

"I know, Mary," he explains bitterly, "*we* are supposed to make sure every system works every time, but somebody saved money by reducing direct labor in Engineering. We lost good men, got behind schedule, and had to release before we were ready."

The factory test engineer cuts in, "Now we have to peak-up all the readings with dozens of test engineers and technicians, on each one of 500 equipments, because Engineering did not get the time and money to stabilize one! . . . Some savings!"

The consequences of "savings" on existing direct labor should be carefully examined.

Tomorrow's direct labor is something else. In the example just cited it would take 10 engineers and 20 technicians, for the life of the contract, to conduct "systems test" on the equipments which Engineering was required to release ahead of schedule.

It would take four engineers only two months to stabilize the circuitry of the original design. The task group, after touching base with the chief engineer and the plant manager, recommends a review of the circuit design, conducted jointly by Design Engineering and Factory Test Engineering.

Where does the money come from for such a review? At the general manager's staff meeting the plant manager graciously offers to transfer enough funds from his "systems test" budget to cover the design study. He makes out like a bandit; the more stable equipments will sail through the factory smoothly. There will be savings in faster product flow as well as the direct labor savings in not hiring the additional test engineers and technicians. Let it not be said that these are not savings. If money is budgeted to go down the drain and we use some of it to plug up the hole and release the rest for other uses, *that* money has been saved.

The chief engineer also makes out. For once he gets the time and money to complete a design, and his department is temporarily protected from thoughtless cutbacks.

Note that faster production flow was a major gain. Faster production flow makes it possible to save money on existing direct labor without taking a single operator off the line, without eliminating a single pay envelope.

How? . . . Say a product is modified so that the savings in labor are spread over several operators and no single operator can be taken off the line; for example, a unit is assembled and inspected in 60 minutes, progressing through 10 stations at 6-minute intervals.

If we can save an aggregate of 10 minutes so that *the same number of operators* move the unit at 5-minute intervals, we will get 12 units an hour instead of 10, saving some 16.6% in the cost of labor per unit *without taking an operator off the line.*

Now, savings in factory direct labor are regularly accomplished by the many industrial engineering skills which organize and methodize production, that is, by the technical personnel who support the work force. One of them should always be included in a value task group, not from the standpoint of savings alone but of product improvement as well.

This is the person who touches base with his friends in the factory. He brings them on board early and they help the task group develop and calculate the savings. The egg is now in their barn and they have the money to incubate it; for example, the Factory-Support Engineering account can take care of design and drafting, and a number of factory accounts exist expressly for the purpose of reducing the cost of manufacturing.

Materials Management Funds. If payroll is one of the two major valves governing the outflow of cash, the cost of purchased materials is the other. This cost is kept in line by the traditional techniques of negotiation, volume buying, new sources, and purchase-value analysis.

In the latter case the supplier often implements the value-analysis recommendations when he realizes that they will improve the value of his product. Such recommendations may come from a value task group in the customer's plant, or the customer may provide the guidance for setting up a task group in the supplier's plant. A combined group from the two plants works very well.

The usual results are an increase in gross margin, which the supplier would like to see as an increase in profit and the customer would like to see as a reduction in price. A little of each is better.

The more vigorous plants in large corporations, and many a progressive young company, choose to go beyond traditional techniques in order to produce better than traditional results. Such dynamic outfits have a plant-wide value program to improve the interchange of information at the beginning of each key phase in the life of their products. Once such a program has proved its worth, it gets its own implementation budget. In a typical multiplant division producing consumer products the return on the value program has been always more than 10 to 1; that is, $10 gained for every dollar spent; BUT, don't rush off to build a plant that manufactures nothing but value analysis. The 10 to 1 reflects a skimming process of the best opportunities, and the dollar spent is spent mainly in communication among people who were paid from other sources. The value added has been the value of better information and faster communication—which often converts fruitless into fruitful effort—much to the benefit of whoever is actually paying for it.

Finance—Where the Funds Are. Think of them as "bean-counters" and you won't get any beans. To get the beans you need you have to understand what the people in Finance really do. Besides, they never counted beans. It was checkers; hence the title chancellor of the exchequer. The exchequer was a hall in Westminster where the royal accountants wrestled with pounds, shillings, and pence in *Roman numerals!* They had inch-wide grooves on their tables along which they slid checkers, which were the "bits" of their computer. But the chancellor of the exchequer did more than count the king's money. He was responsible for collecting it as well.

The modern financial manager starts out by procuring the money to run the business and by educating his fellow executives on the time-value of money. Then he develops methods for finding out and reporting where the money is doing the most good and where it is being wasted. This leads to recommending investments and controlling waste. We do not have to go much further to illustrate the common interests between Finance and Value Analysis.

On the one hand, Finance can help us find the money to implement our proposals. Being familiar with all budgets, and with the cost of waste, they can point out who will benefit most from a proposal to eliminate waste in any form. Since they depend—even more than the rest of us—on billings for their well-being, they can see the sense of investing some money to make the product more appealing to the customer. In short, the financial manager and his subordinates are natural allies of value analysis.

On the other hand, Finance can prevent much Love's Labor Lost on "savings" which turn out to be not savings at all. We should take the trouble to go to Finance *first* to find out where the pay dirt is.

PLAN OF IMPLEMENTATION

Everything described in this chapter so far refers to getting something done. We have considered WHAT must be done, WHO will do it, HOW the funds will be charged, and WHERE and WHEN it is to be done. The best way to align these elements is in a Gantt chart, modified to include the WHO, WHERE, and HOW as shown in Figure 9-1.

The workshop effort culminates in a presentation before the very people who will have to authorize or implement the recommendations. Having been briefed together, the implementers will not have to explain to each other what they are trying to do and why they are trying to do it. It is also easier and faster to present the plan to all the participants together than it is to present it, and explain it, individually.

THE PRESENTATION TO MANAGEMENT

The formal value workshop ends with a presentation before the plant manager, his staff, and managers responsible for the products analyzed. Most of these people are mature business executives who should not be burdened with too much detail. They want bottom-line answers to the following questions.

What will your project buy us?
What will it cost us?
What are the risks?

The spokesman for each value task group has to arrive at the required bottom-line answers by the shortest, most straightforward path, explaining who carried out the analysis, what it was about, what the problems or opportunities were, and what improvements were proposed. Figures 8-8 through 8-14 show typical flip charts which identify a team and its task. The particular skills of the members are given in order to show the multidiscipline nature of the group.

Generally the "TEAM" chart or slide should list the members, their occupation in the plant, and their particular chore in the task group, such as *captain, recorder, spokesman.*

Proposed improvements should be described without too much detail, generally in one or two charts. All further explanation and substantiation can be kept in back-up sheets. At this point the spokesman for a team begins to answer the bottom-line questions.

What Will It Buy Us?

Because of the fresh information and rapid communication available to a value-analysis team, more opportunities for improvement are discovered than the team was initially looking for. Even though the initial goal may have been cost reduction, a joint look at the product by the key specialists almost always yields additional benefits. Conversely, improving performance or appearance often reduces cost simply because inefficiency and ugliness are expensive.

The following check list should serve to make sure that none of the benefits is overlooked:

Improved delivery
Improved production cycles
Improved performance
Improved reliability
Better quality
Reduced weight
More suitable size or shape
Better appearance

Dollar savings usually have the greatest impact because they represent dollars, here and now, that can be kept from going out the door. Savings should be estimated conservatively.

What Will It Cost Us?

Carrying out any change steals time from the regular operation. The costs listed below should be generously calculated:

Loss of product during changeover
Schedule delay
Scrap or rework
New tools
Additional or more costly materials
Engineering cost to implement
Manufacturing cost to implement

Cost of implementation combines all the applicable costs mentioned

above and any other costs the team may know about. In the presentation, however, only the total is usually given. The "SAVINGS" chart
may be as simple as this.

GROSS SAVINGS	$104,000
COST TO IMPLEMENT	8,000
NET SAVINGS	$ 96,000

OR 32% OF $300,000 INITIAL COST
PAY-OFF PERIOD   One Month

What Are the Risks?

Beginning with performance of the product's function, customer satisfaction is the area to be guarded most carefully, for performance problems always seem to come as a surprise. Procurement and manufacturing
problems are reasonably predictable, but market hazards, related as they
are to the appearance, convenience, or image of the product, are the
trickiest and most deceptive kind of risk that the task group dares to
incur. The customer does not have to tell us, "I don't like it," or "It
doesn't feel right." He buys something he *does* like, even if ours works
better and costs less. When esteem value is in danger, the task group
needs all the help it can get from Styling, consumer surveys, trade shows,
distributor conferences, the analytic tools of marketing research, and
from all the hidden powers of Heaven and Hell.

Here is the check list to be used in identifying risks:

Performance problems
Procurement problems
Manufacturing problems
Market hazards
Other hazards

Table 9-1   Example of Risks (an Old Design)

Performance problems	None
Procurement problems	None
Manufacturing problems	$ 2,500
Market hazards	None
Other hazards (reserve)	$ 1,000
Expected risk	$ 3,500
Expected savings (1 yr)	$96,000
Expected net gain	$92,500

Table 9-2   Example of Risks (a New Design)

Performance problems	$ 10,300
Procurement problems	None
Manufacturing problems	None
Market hazards	Moderate
Other hazards (reserve)	$ 5,000
Expected risk	$ 15,300
Expected gross profit (1 yr)	$210,000
Expected net profit	$194,700

The information leading to the appraisal of risk is developed as follows:

PERFORMANCE PROBLEMS. None anticipated. Prototype has been fully tested. New components are more reliable than old ones (Table 9-1) or . . .

The task group is working with the design engineer on a new design (Table 9-2). Performance problems are anticipated as the result of line-voltage irregularity. Chances of line voltage becoming a problem are about 20%; cost of the fix would be $1500, and added cost per unit would be $50. Expected risk: 20% of $1500 = $300 once; 20% of $50 = $10 × 1000 units in 12 months = $10,000. Total expected risk: $10,300 first 12 months.

PROCUREMENT PROBLEMS. None.

MANUFACTURING PROBLEMS. Chances of requiring an additional press are only 10%; cost would be $25,000. Manufacturing expected risk $2500 nonrecurring (Table 9-1).

MARKET HAZARDS. None. Styling approves the simpler form factor and satin finish; doing away with the gingerbread of older designs should please the customers. Marketing agrees.

OTHER HAZARDS. With a brand new design—pushing the state of the art—we should expect the unexpected: about $5000 worth of surprise problems. On the old design $1000 reserve should be enough.

The figures given above for the various risks are educated guesses based on experience. Often we have nothing more than our intuition and judgment as a basis for appraising risk. The thing to do, then, is to level with management and use such intuitive terms as *high, moderate,* and *low.* The term *moderate* next to *Market hazards* in Table 9-2 is an example. It refers to the higher sales price resulting from an increase in cost if the manufacturing problem *does* materialize.

The written report should be tailor-made for the particular executives who have authority to accept the project, provide the needed resources, and order implementation. Before preparing the draft the team should answer these questions:

Who will authorize the action?
Who will implement it?
Who should see the draft?
Who will verify the figures?

The Five Parts of the Report

The first two parts describe the project and the team. They can be copies of the flip charts. As a matter of fact the flip charts are often photographed and included in the report. The third part, a description of the proposed improvements, must include what is missing from the flip charts—complete how-to-do-it information. The fourth part, really a cost accounting effort, provides all the cost, savings, and net gain figures. It presents the project as a *business* venture and of course includes the appraisal of risk, the pay-off period, and the break-even point. The fifth part is the plan of implementation. Here is an outline for the written report:

1. The project
2. The team
3. Recommendations
4. Cost, risk, and net gains
5. Plan of implementation

As mentioned earlier, the key points, and only the key points, of this report are highlighted by the spokesman of each team in the verbal presentation. One way to keep this presentation convincing and short is to let the best speaker on each team do the speaking. Another way is to avoid the individual introduction of team members. Instead of calling out the person's name and having him, or her, stand up, the team members follow the spokesman to the front of the room and sit facing the audience. By a gesture the spokesman implies that there is a one-to-one correspondence between names on the chart and team members in their places.

Other advantages of having the team near the spokesman is that they can give him more direct support during his presentation. He can refer to them for specific numbers and they can hold pieces of equipment in their laps and do the show part of his show-and-tell bit.

PUBLIC SPEAKING BEFORE YOUR BOSSES

In the same way that the law of contracts requires two or more parties, an agreement, and some good and valuable consideration, the law of communication requires two or more points, a message, and useful information. The first point is the point of transmission.

The Transmitter

Whether it is a nerve cell "firing" a message to its neighbor, an Indian making smoke signals on a hilltop, or a TV station, the transmitter expends resources. It takes power to send a message. Because resources used in communication can trigger action of far greater magnitude, they should be invested generously. In short, if you are a nerve cell hollering "ouch," an Indian making smoke signals, or a radio transmitter sending the human voice from the surface of the moon, you should discharge enough energy, put enough wood in the fire, or use enough power to be heard.

Full Capability of the Human Voice. In public speaking—unless you are a heart patient and no brandy is available—don't save your voice. Be a powerful transmitter. Let the words roll out, loud and clear, to the farthest corner of the room, and if they rumble on out the door and along the halls let people say, "Something important is going on in there. Somebody is really telling them. Who *is* he?"

How to do it? Throw in the resources, starting with the cheapest—air. Fill your lungs, then use enough energy to drive the air through the resonant voice box of your entire upper body.

Avoid Strain. Your heart and vocal cords are the only items of capital equipment that you have to watch. You might say, "My heart? Ha. I'm fit as a fiddle," but haven't you seen perfectly healthy football players, interviewed *before* the game, running out of breath?

Public speaking makes extra work for the heart and at the same time robs it of life-giving oxygen; so pause and take a deep breath between sentences. The audience will think you are trying to emphasize a point. Fine. Be sure you emphasize enough not to run out of breath in the middle of your talk.

Strain on the vocal cords depends on tension and frequency of vibration. The next time you stop at a music store ask them how many "high E" strings they sell for each of the "low E's." To protect your vocal cords and yet process more vibrating air through the sound box, you are better off speaking in the lower registers in a rich voice.

Among the early Indo-Europeans, whose natural eloquence is revealed to us in the Old Norse Eddas, Homeric poems, and Vedic hymns, the Celts had a particular veneration for the power of the human voice. Compare the stately roll of words, "By the r-r-r-ight, quick, mar-r-r-ch" thundered by a Highland Scot pipe major with the strangled scream that passes for a military command among his English brethren. Let your voice roll out deeply from the chest so that it sounds more like the tones of a great organ than the arguments of Donald Duck.

More than Audio. If voices alone carried the message, the president of the United States would not address the nation on TV. The attitude, poise, and posture of a speaker create a vital first impression. His very walk to center stage should tell the audience that he knows what he is doing, that he has something to say, and that he is well prepared to say it.

Spare the Corn. If you are a comedian hired to entertain the audience, a preacher without a sermon, or a politician without a program, a joke is in order. But if you *have* something to say, start saying it. It may be that humor is the best way to put your initial point across. Fine. Humor is a great economizer, often utilizing the audience's previous knowledge and experience to let them see in a flash what might take many sentences to explain. If you can carry it off, by all means use humor to illustrate your points, but don't, unless you are before a night-club audience, drag a joke in for its own sake.

The Importance of Posture. If you slink up to the protection of the podium, crawl behind it, and with downcast eyes start mumbling in a dreary monotone, the victims will turn off their receivers. On the other hand, if you face the audience with the aplomb of a man who has a few well-prepared remarks with which to accept the Nobel Prize, you will have them eating out of your hand. Your posture should reveal confidence and pride in what you and your teammates have accomplished. Stand tall and speak loud and clear.

The Receivers

Innumerable combinations are possible among transmitters and receivers. We will now look at them in order of decreasing difficulty. The worst combination is the one in which a single member of an extended family is the receiver and all the other members transmit to him simultaneously. Not much easier, although the most common of all, is the supposedly alternate communication between two people. It is peppered by interruptions, simultaneous transmission, and side-tracking remarks. The multilateral communication of meetings is easier because most members actually listen to the speaker; if not, the chairman says, "Let's have only one meeting." The easiest of all communication is when *you* are the speaker and all the others are listeners. That is what happens when you have an audience.

They are yours from the start, but you can easily lose them. Even if you talk loud enough and avoid irrelevant jokes, a droning monotone, and too much detail, you can still lose them through repetition. They will tell each other "He said that before—at least five times" and they will turn you off. You can also lose them by talking to the flip charts, to the blackboard, or to your notes instead of to the audience.

You can hold your audience with your eyes as well as with your voice. If you must give your talk from notes or from the flip charts, fine, look at the notes or the chart, then face the audience and *tell* them what you have read instead of mumbling to them *as* you read.

Why read the flip charts or view graphs? If the writing is too small, too sloppy, or too cluttered for the audience to read, it should not be there at all. If it is neat and legible, *they* can read it. It is all right to read the one bottom-line figure on a chart that you want to emphasize, but there is no excuse, outside of first and second grades in elementary school, to stand up there and read word-for-word and number-for-number what is on the chart for everybody to see.

The Message

The message is what you say, write, or do to convey information. In presentations to management in a value-analysis workshop the capacity of the message is governed by the time available to each team, usually 8 to 12 minutes—often less, and seldom more.

Let us say that the information you would like to present would require a 20-minute message but you are ruthlessly forced to boil it down to 12 minutes. Do you eliminate the most significant fact? No, you eliminate the least meaningful parts of the message. You end up by throwing away the oyster shell, eating the oyster, and presenting management with the pearl.

Of course, it is not really so simple as that. If you did not think that the shell and the oyster were important, you would not have included them in the first place. They served a purpose, showing how the pearl had come about.

This would be excellent reasoning if you were an experienced pearl diver teaching your little boy the origin of pearls. But you are talking to management, to veritable giants among pearl divers. They already know about oysters and shells—all they want is the pearl. The more experienced your audience, the less you have to tell them, and the fewer details they will put up with.

Usefulness of Information. Information that displaces or obscures more useful information is simply noise. To be useful information must help the recipient select a future course of action: (a) by revealing new options; (b) by providing the means of comparing available options; or (c) by making certain options exclusively available to the recipient or his company.

To get more information into the time available give the audience one visual message while you are giving them a different oral message and have your teammates illustrate the oral message with the pieces of equipment you are talking about.

If the information is relevant and interesting, the audience will absorb it. Remember, an adult can enjoy a three-ring circus while eating popcorn and answering children's questions.

Practice. Dry-run your talk before your teammates: first for timing, without interruptions, then for voice, posture, and projection. Each talk should then be dry-run before the whole group.

SUMMARY

VOLUME.   Use enough air. Talk from the chest in your deepest voice; let the words roll out loud and clear.

POSTURE.   Face the audience squarely. Stand up straight. Hold your head high.

DIRECTION.   Talk to the audience, not to the charts, the blackboard, or your notes.

CONCISENESS.   Throw away the oyster shell, eat the oyster, and give your audience the pearl.

So far, all you have done is tell management what is at the end of the rainbow. Now you have to show them how to get there. Near the end of your talk, or in the actual windup, you can tell them about the plan of implementation. You haven't got the time to outline the plan, but you can tell them that they will receive a draft of it. The chart in Figure 9-1 shows a condensed way of presenting such a plan.

A STATE DOCUMENT OF IMPLEMENTATION

His Britannic Majesty acknowledges the said United States to be free, sovereign and independent States . . . There shall be a firm and perpetual peace between his Britannic Majesty and the said States, and between the subjects of the one and the citizens of the other . . .

Done at Paris, this third day of September, in the year of our Lord one thousand seven hundred and eighty-three.

For his Britannic Majesty   D. Hartley

For the United States of America   John Adams
                                                             B. Franklin
                                                             John Jay

X

DIRECTING THE OPERATION

Who can question the money-making power of such a product as Angostura Bitters, a product that can be found in the uttermost corners of the earth? Yet Angostura Bitters was developed for a much humbler destiny than it finally fulfilled.

The same is true of value analysis. The original need was for a fresh approach to cost reduction, just as the original need for Angostura Bitters was to cure bellyaches among the Indians of the Orinoco.

Today, there are innumerable cures for bellyaches but only one Angostura Bitters, which outsells both cures for bellyaches and all other aromatic bitters put together. There are also innumerable cost-reduction programs, both useful and necessary, but value analysis has revealed a capability far beyond cost reduction.

The town of Angostura now calls itself Ciudad Bolivar, in honor of *El Libertador,* whose army suffered its Valley Forge in this deep jungle hideout.

Among the few foreign volunteer forces that remained loyal to Simon Bolivar in those days of adversity were the Brigade of St. Patrick and the Brigade of St. Andrew, made up of Irishmen and Scotsmen, respectively. Surgeon general to this force was Dr. J. G. B. Siegert who learned of Angostura Bitters from his colleagues, the Indian medicine men. He prescribed it for the stomach—four teaspoons in the soup.

The Irish and Scot grenadiers soon found liquids truly worthy of the wonderful bitters. By 1830 Dr. Siegert was exporting the bitters from

Angostura, and today millions of people who have never heard of the
South American Wars of Independence, and who have no bellyaches,
happily buy Angostura Bitters, for it meets a higher need—mankind's
insatiable need for improvement.

Today's value analysis meets more urgent needs than simple cost
reduction; it meets the needs for current, relevant, and accurate information, fast communication, and smooth convergence of disciplines on a
given product.

THE BITTERS DO MOST GOOD UPSTREAM

Advanced Research and Development

A joint value task group, for example, can do wonders for the research
laboratories of a manufacturing company and a chemical company when
the first is praying for somebody to develop a particular chemical and
the second is praying that somebody can find a use for the new chemicals
they are developing.

Results of such task-group meetings are that the researchers, in both
houses, no longer have to grope in the dark. The one has a future supplier and the other a future market. The law departments of both firms
must establish safeguards to make sure that whatever discoveries are
made do not exclude competition so that nobody will go to jail.

New Product Development

This type of task group works under the direction of a division's New
Product Department. It represents the various skill centers, the cus-
todians of resources and facilities, and the Purchasing, Marketing, and
Styling functions. The division and corporate value analysts do all the
dog work. They prepare the meetings, present the method, gather additional information, and guide the discussion.

The rest of the product cycle is discussed elsewhere. One general rule
applies. The value program offers the tools and the opportunity for effective interaction among specialists, but it does not, as a rule, attempt
to do their work for them, and it should never usurp their authority or
invade their area of responsibility.

"Justifying" Upstream Value Analysis

When the sole purpose and only output of a business method is cost
savings, everything it does has to be justified in terms of savings; but

when a method contributes perceptibly to improved performance it justifies its existence on the basis of better products and faster production cycles.

Easier Said Than Done? Not really; not if Research and Engineering explain that they can release a product earlier, with less chance of procurement and manufacturing problems, *provided* a value task group works on it. The advantages of that "provided" can be measured in dollars and cents.

Who pays for the task-group members?

Their bosses do, just as they were paying for them all along and for the same work too, except that it is done earlier and with surer information. Moreover, these bosses get advance information on what is entering the pipeline and have an opportunity to contribute their specialized inputs early enough to do some good.

The question arises, of course, how to get Engineering to ask for value analysis? Not by persuasion, exhortation, or pressure but by an initial controlled experiment, followed by one or two others to ensure repeatability. The second and third efforts flow very smoothly.

How about the first effort? Let's say you make your pitch to the engineering leaders. Show them what the method has done for others. If you can get such a beneficiary, even if it is from another company, let him tell them. Show them numbers, particularly in terms of improved schedules and greater producibility. From the audience's standpoint the biggest advantage is that the multidiscipline approach releases engineering time for engineering tasks.

They applaud at the end of the presentation; then, dead silence. Now, acknowledge the possible objections: their group may not be ready for such an experiment; departure from the traditional way does involve some risk; professionals, such as themselves, may not want to rub elbows with persons of lower cultural and social status, and so forth.

Suddenly a mean-looking curmudgeon, off in a corner of the room, growls, "I think I'll give it a try."

Then break your back for him!

HOW TO INTRODUCE A NEW PROGRAM

> . . . *the innovator has for enemies all those who derived advantages from the old order* . . .

Niccolo Machiavelli, *The Prince,* Chapter 6

Doing the corporation a lot of good at the expense of one particular department does not create enthusiasm in that department. They are afraid of having their weaknesses revealed and of having to do more work without getting the time to do it in. You will have to allay these fears. Your program does not have to hurt anybody. As a matter of fact, one of the secrets of most successful value programs is that they save time, save money, and improve the product *without hurting anybody*. But all this comes later, after you have chosen the type of program most suitable for your target plant or division and after its boss has said yes.

Choosing the Approach

There are three basic approaches: the independent value-analysis group, the workshop, and the train-everybody approach. They are not mutually exclusive.

The independent value-analysis group is made up of *doing* analysts. These men look at a product and apply the sequence of value analysis steps to define the function, place a value on it, and to find a better way of accomplishing that function. The problem here is that, since the group itself does not originate a product, design it, or manufacture it, its work must, of necessity, constitute an additional step and some duplication of effort.

Suppose a product is brought by the product manager to an independent value-analysis group. The value specialists look at the product and recommend a number of improvements. The product manager then puts the product on the chief engineer's desk, lays the recommendations before him, and says, "Please do thus and so."

Does it get done? Alas! No. The chief engineer defends his operation and points to the innumerable reasons why the value recommendations cannot be put into effect.

And he may be right. He has been left out of the improvement task; so have his experience and his knowledge of the product. The truth of the matter is that the second-look independent approach provokes conflict between the second lookers and the originators of the product. The sad result is that many value recommendations are simply not implemented. The few that *are* implemented often embarrass the people responsible for the product, and thus create hostility to value analysis.

What's the answer?

A Cooperative Value-Analysis Group. Such a group is independent in the sense that it is virtually autonomous and can offer technically independent opinions. It is cooperative in the sense that it identifies the people responsible for a design, brings them on board early, and makes its recommendations to them and not to their bosses.

The Multidiscipline Approach. This is the method in which the team members represent those skills that contribute most directly to a product's value. This method should begin at the earliest stage of the design concept. The engineer responsible for the design calls the value analyst and asks him to organize a task group that will meet two or three times a week.

One advantage of this approach is worth noting here. The work is done by people who are paid to do the same thing at their desks. It does not cost any more when they do it together. Yet everybody wants to know how they get paid.

"I can't afford value analysis," a young engineer once told me, "no budget for it."

He's raising a roadblock, I thought. Through bitter experience we have learned that the "roadblocks" raised against a proposal simply mean that the proposal is either not good for the person raising the roadblock or it has not been presented right.

"Look," I told the engineer. "I am not talking about a bunch of little helpers puttering around on their own. I am offering you a personal staff, representing disciplines other than yours; seasoned professionals, each an expert in his field, and all answerable to you. They will work for you and not around, and you—"

He threw up his hands. "I *know* Carlos, but I just haven't got the money."

"Do you ask Santa Claus how he pays for the toys?"

"You mean, *you* can—?"

"Leave it to me."

This happened some years ago. I started out by tackling the division's somewhat authoritarian purchasing agent.

"Sir," I quavered, "we need one of your buyers, say Slim, for a value-analysis task group on—"

"No!" he thundered. "Do you realize that when the paper piles up on a buyer's desk," the growl became a roar, "*production stops!* . . . ? Why, Slim is my best electrical buyer"; then, in a crisp military tone, "request denied, dismissed!"

I did a smart about face and marched out. Wrong approach. Delayed the task group a week and I know why, too; I had entered his office like a suppliant asking for alms. Besides, it was 11:30 in the morning and we were both hungry.

"Feed him," I told my boss, "give him a big, leisurely lunch. Don't ask him to give me an appointment, tell him to be in his office at one-thirty because I have something important to tell him."

As I marched through the Purchasing Department I saw it as a battlefield. I could almost hear the rolling of drums. Suddenly I was in the Ogre's den, past his terrified girl.

Sitting on the edge of his desk, I shoved my face up to his. "Do you know what Engineering is up to?" I asked in alarm. "What they are going to do on that big classified contract? Do you know what's going to hit you two months from now?"

He turned pale. "No. They never tell me anything," he whined. "They want the stuff yesterday and tell me about it tomorrow. I don't even know what they have in mind. At the last minute they'll call for stuff that can't be procured!"

"I'll help you infiltrate," I reassured him. "I can sneak one of your buyers up there, say Slim; then he can make sure that what they specify *is* procurable. You'll know the requirements before the requisitions are written. You'll have advance—"

He was pressing buttons on his intercom. "Get Slim in here at once!"

Slim appeared at the door, briefcase in hand—I had told him to get ready.

"Go with Carlos," he was told. "Do anything he tells you to. Never mind going back to your desk. I'll have somebody cover for you."

In the factory it was easier. "How would you like to have a manu-facturing engineer work with the designers? He can make sure everything is producible and you can begin tooling earlier."

"Sure," was the cheerful answer. "You see those empty clipboards? They would normally have problem sheets on them, but we sat in on the value task group. For once all the parts went together."

The controller was the most willing of all. "Before-the-fact control? Know what things are going to cost before they cost? I'll have somebody there in the morning."

The truth of the matter is that both task-group members and partici-pating departments come out ahead. Team members establish valuable contacts across the entire organization. They learn to work with people of different disciplines and they usually become more promotable.

Each participating department exerts timely influence on decisions that will affect it later and it gains advance information on the product.

The general training approach strives to put as many people as possi-ble through a value training program in the hope that value analysis on real business projects will be done by the trainees when they get back to their regular work. This approach gains the benefits of multidiscipline enrichment during the training seminar, but it is like having a company basketball team that practices after hours. The training does contribute to their other duties. They know the other team members and cooperate more closely with them and they invariably make a goal when they throw paper cups into the wastebasket; but basketball *and* value analysis are best played by the team *together*.

Many of the projects in a training seminar are real-life projects which generate real savings, but they are especially selected for training purposes in order to give the participants a success experience.

Such a training project was chosen by a gunner's mate in the Colombian navy when his aunt asked him to train a young cousin to be less shy with girls. The cousin was a dropout from a seminary. He had dropped out straight into the marine corps and, after boot training, had been assigned to my destroyer. It did my heart good to watch the gunner's mate taking his young cousin, and a pretty girl, for a sail in one of the ship's 16-foot dinghies.

Shortly after dark he returned without the young people, but with their clothes. "Don't look so shocked, sir" he told me. "They are wearing bathing suits. I left them alone on one of the outer islands. The night breeze is cold; he's going to have to talk to her."

Whether he talked or not, I do not know. But they had a success experience, all right! We worried about it for 17 days; yet the boy did not learn courtship techniques suitable for general application.

If it uses projects more realistic than the one just described, the training approach has considerable merit. The training seminars provide from 32 to 80 hours of intensive training over a period of one or two weeks.

Many a company has started its value program with a one- or two-week seminar put on by a consulting firm, usually with a follow-on contract to make sure the program does not die.

Organizations for management training also conduct orientation seminars to give managers a general idea of what value analysis can do for them.

In a training seminar, after a few orientation lectures, the students are grouped into teams which then put a number of training projects through the value-analysis job plan. They do learn, provided the projects present a real challenge—the girl should at least be wearing clothes.

I have personally found that real-life business projects in a profit-oriented workshop provide realistic training, enough money to pay for implementation, and something for the company treasury.

Let us say that you have chosen your approach. Your next move is . . .

GETTING APPROVAL

If you tell a general manager that you want to make a presentation to his staff in order to launch a value-analysis program he may answer you politely with a number of evasions, but he will be thinking that he is not ready to launch any such program. What he needs to know is the minimum work, expense, and risk necessary for him to *appraise* the program. So you simply ask him for 45 minutes to explain the program to him and to his staff, with no further commitment on his part. Nine out of 10 such presentations are successful.

The Introductory Talk

Tell the group what value analysis can do for them as managers. Mention the opportunity for savings in elapsed time, in the early interchange of information, and in improved communications at the working level. This more or less tells them what value analysis can do for them. The next thing is to tell them what it will *not do to* them. It will not reveal previous mistakes. It will not show up mismanagement; it is not meant as an investigative technique of how not to do things but as a positive forward-looking method for doing them better.

The talk should wind up with a request for making a similar talk to the next lower level of management. This is the technique of the little boy in the fairy tale who was granted three wishes and on his third wish asked for three more.

The Planning Session

This presentation, which is your second 45-minute talk, is before a much larger audience. This audience will be made up of the staffs of the people present at the first talk. At this time you should explain how value analysis can contribute to better communication among their subordinates, a smoother flow of information, and a planned look at a product instead of fighting emergencies as they come. By this time the audience should be nodding approval, not because of magic words or tricky salesmanship but because value analysis *does* offer them something that is good for them as individuals as well as for their departments. If all goes well—and in most instances it does at this stage—people in the audience will begin asking about products suitable for analysis and about the people who will do the work.

CONDUCTING THE VALUE-ANALYSIS WORKSHOP

A typical value workshop will include anywhere from 3 to 10 task groups of 4 to 6 men each. The work schedule is arranged in half-day sessions, each preceded by a discussion on the applicable principle or task that will occupy the session. The half-day sessions can run together; that is, you can have four working days all in one week, eight half days spread over two weeks, or a longer period, say two full days a week or even two half days a week. I have found the most practical method to be 48 hours of work distributed over two or three weeks, generally in half-day periods. This leaves the people time to go back to their regular shops, read their mail, keep the job going, and acquire the necessary information for the next session.

The task groups or teams are selected with specific products or services in mind. First we select the products or services to be analyzed and then tailor the task groups to suit these products or services, making sure that the major disciplines that contribute to product value are represented. Value workshops run anywhere from a minimum of 32 hours to 120 hours, which is about the longest time any organization can permit that many key people to be away from their jobs. The biggest training advantage of value workshops is the preparation of people for service in ad hoc task groups which can be set up at any time and may well take longer than 120 working hours.

Schedule, Goals, and Pressure

One of the advantages of value analysis is its quick response and the momentum gained from enthusiastically working under pressure. For this reason I believe that the task group should always have a deadline as well as specific goals, both in savings and in product improvement.

Vendor Participation

I have found vendors most cooperative. Level with them. To avoid disappointing many vendors and boring your own people make sure that vendors are selected to suit specific products or problems. Don't bring vendors in just to show their wares. Make sure you have a specific application in mind.

Vendors can prove amazingly helpful if motivated properly. In an economy as rich as ours the old, harsh ways of motivating vendors do not work but becoming a better customer does.

When we run a value-analysis workshop at a vendors' plant, for example, we do not question his profit. If cost is reduced, we never claim the entire reduction. We think a portion of it should go into the vendor's profit in order to motivate him and other vendors to ask us for more value analysis. After all, 60% of something is better than 100% of nothing. I have never found a "vendor's day" practical in the workshop. Some vendors should be invited from the beginning; others may receive a particular task group in their own plants. As a rule, each task group maintains its own particular vendor relations through the purchasing representative in the support group.

The support group is made up of design engineers, product managers, cost estimators, marketing and styling specialists, and at least one buyer and one manufacturing representative. These are usually the line people responsible for the products being analyzed or their representatives. They are kept fully informed and they provide information. They also brief each team on the product that the team will analyze.

Touching Base with the Support Group

You have to pave the way for approval as soon as you find a possible improvement in someone else's area. Bring that someone else on board early. He probably had the idea in the back of his mind all along. Let's hope he did. If it's his idea, it is much more likely to be implemented. You will get no credit for it, but the word will get around that wherever you go other people save time and money. This person may or may not be a member of the support group. If not, someone in the support group can lead you to him.

Beware of Self-Righteousness. Some actual and many apparent departmental shortcomings will appear during the workshop. Caution! Remember that value analysis is a fast response method which has to work with people as they are and with the industrial organization as it is. Of course, there is a need for people to look at industry as it should be and to value-analyze the whole plant as described at the end of Chapter 8, but when we are out to improve a given product it is easier to do it in the plant *as it is.* The value-analysis task group, therefore, cannot afford to present recommendations in a way that will make a key department look bad. Such recommendations are simply not implemented.

Preimplementation Strategy. In the same way that a rider canters up to let his horse examine the fence to be jumped before attempting to jump it, the experienced value analyst today backs into the analysis task from the goal of implementation. Pertinent questions at this time are these: "If we come up with some good ideas, who will have to approve them? Who will have to do the work? Who will we be overloading? Who will be happy to do what we want?" You can imagine how much advance planning must be done by the group in order to direct their efforts to fruitful areas.

Table 10-1   A Road Map

---

1. SELLING THE VALUE-ANALYSIS WORKSHOP

The gains in time and dollars usually sell the program purely on the basis of results—measurable, short-range results that are not obtained at the expense of future gains. This is a low-risk approach on two counts. First, management can throw out the program if it fails to produce good results soon, and, second, management is committing only a modest portion of today's funds. An initial offer should be made to the client's plant manager in his capacity of entrepreneur, thus giving him the chance to risk departing from his routine operation for the sake of greater than routine gains. If he is willing to take the chance, the plant manager will then get his staff together and let the value specialist present the opportunity to them. The presentation goes something like this.

1.1 Discuss the proposed seminar with the general plant manager and his staff
    1.1.1   What will it buy them in terms of
        (a) improved rates and schedules, faster turnover
        (b) improved yield
        (c) reduced cost
    1.1.2   What will it cost them in terms of
        (a) management time and effort
        (b) number of people tied up
        (c) dollar cost
    1.1.3   Timing and duration of concentrated effort with respect to
        (a) general operating schedule
        (b) trade shows, general sales meetings, product exhibits, conventions
        (c) start of production runs, product changeover, product phaseout
        (d) plant shutdown, other vacations, year-end break
        (e) inventory taking, audits, movement, or rearrangement of facilities

This is not to say that any of the events listed need conflict with a value workshop. Some of them could provide the best time when the desired people are available, but the value analyst must make certain that the workshop is timed so that the recommendations can flow into the plant's general program with the highest likelihood of implementation. Other items to be discussed with the plant manager and his staff are the following:

    1.1.4   General nature and approximate number of projects
    1.1.5   Number, level, and specialties of the participants
        (a) in the task groups
        (b) in the support group

Table 10-1 *(Continued)*
2. PREPARING THE VALUE-ANALYSIS WORKSHOP

The general manager indicates the department that will be responsible for administering the workshop. A person of stature within that department is designated workshop administrator to work with the value specialist in the following areas:

2.1 Preliminary schedule
    2.1.1 Firm dates
    2.1.2 Instructor's schedule and syllabus
    2.1.3 Types and number of project recommended
    2.1.4 Goals in terms of faster product turnover, improved yield, and reduced cost
    2.1.5 Workshop budget
    2.1.6 Request for participants
    2.1.7 Request for facilities
2.2 Selecting facilities and vendor support
    2.2.1 Classroom and work space in the plant, hotel, motel, club, or school building
    2.2.2 Furniture, desk calculators, telephones, and supplies
    2.2.3 Number and type of vendors to be invited
2.3 Negotiating, coordinating, and distributing final schedule
2.4 Selecting the supporting group to cover the following aspects of the anticipated projects:
    2.4.1 Business background and history
    2.4.2 Profit status
    2.4.3 Present and future market
    2.4.4 Engineering
    2.4.5 Materials
    2.4.6 Manufacturing
    2.4.7 Time schedules
2.5 Working with the supporting group
    2.5.1 Follow-up on
        (a) invitation assignment of participants
        (b) invitation of vendors
        (c) facilities
        (d) funding
    2.5.2 Discuss worshop projects with the general manager on basis of business strategy
    2.5.3 Determine the minimum information required
    2.5.4 Get the support group to start gathering this information
    2.5.5 Verify goals in terms of
        (a) acceptable risks
        (b) expected gain
        (c) deadlines
    2.5.6 Select the most promising projects on the basis of

Table 10-1 *(Continued)*

      (a) timeliness
      (b) potential savings
      (c) likelihood of implementation
      (d) opportunities for faster turnover and improved sales
2.5.7   List constraints and investigate them for
      (a) authenticity
      (b) validity
      (c) current applicability
2.5.8   Summarize real constraints
2.5.9   Invite vendors whose specialty is related to the projects selected
2.5.10 Invite the in-house specialists whose skills bear on the projects selected.

3. ORGANIZING THE VALUE-ANALYSIS WORKSHOP

The workshop administrator with the advice of the value specialist carries out these tasks.

3.1 Assign personnel to achieve the following:
    3.1.1   The necessary combination of skills
    3.1.2   Objectivity through compensating biases
    3.1.3   Representation of departmental interests
3.2 Appraise final workshop resources to determine extent of effort
    3.2.1   Personnel actually present
    3.2.2   Time finally allowed—duration of program
    3.2.3   Information actually available
    3.2.4   Facilities assigned and inspected
    3.2.5   Funds allocated and available
3.3 Review program of instruction
    3.3.1   Check instructor's acceptance and mail them the final schedule
    3.3.2   Review notes and critiques from last two workshops
    3.3.3   Verify availability and working order of telephones, desk calculators, slide or movie projectors and other visual aids
    3.3.4   Verify availability of cost-standard books, supplies, flip charts, and materials for final presentation
3.4 Arrange transportation, housing, and food
    3.4.1   Send instructions on how to get there
    3.4.2   Make hotel and car-rental reservations
    3.4.3   Make arrangements for luncheons and coffee breaks
    3.4.4   Arrange welcoming talk by host department where applicable
3.5 Prepare for presentation to management at end of the value workshop
    3.5.1   Set date and time
    3.5.2   Invite the general plant manager and his staff
    3.5.3   Other members of management
    3.5.4   The supporting group
    3.5.5   Staff guests and visitors if desired
    3.5.6   Make preliminary publicity arrangements

Table 10-2   Outline for a Value-Analysis Workshop

---

1. INTRODUCTION

1.1 Opening remarks
1.2 Fundamentals of value analysis
1.3 Plan of the workshop
1.4 Organization of task groups
1.5 Assignment of projects
1.6 Announcements

2. INFORMATION PHASE I

Appraise the task and resources of your task group.

2.1 What is the product to be analyzed—part, model, or contract number and brief description?
2.2 What does it cost?
    2.2.1 Cost per unit
    2.2.2 Quantities involved
    2.2.3 Cost breakdown—identify the elements of cost

3. INFORMATION PHASE II

Examine the data.

3.1 Verify requirements and constraints
3.2 Validate tolerances and allowances
3.3 Identify additional information required

4. ANALYTIC PHASE

4.1 Analyze the project for concentration of effort
    4.1.1 Apply the principle of selectivity in directing effort to the most promising areas
    4.1.2 Plan the expenditure of time and effort in relation to workshop skills, supporting group skills, vendor participation
4.2 What does it do? Define the function
    4.2.1 Pinpoint the function's most important aspect, using a verb and a noun
    4.2.2 Ask "why" to raise definitions to higher levels of abstraction
    4.2.3 Ask "how" to reduce definitions to lower levels of abstraction
    4.2.4 Consider one function at a time
    4.2.5 Search for measurable properties
    4.2.6 Define the function at the highest level of abstraction within the scope of the company's business
4.3 Primary, secondary, and equally important functions
    4.3.1 Identify independent and interdependent functions
    4.3.2 Determine the relative importance of the various functions

Table 10-2   *(Continued)*

4.4   What *should* it do?

    4.4.1   Consider improving the function

    4.4.2   Consider eliminating the need for the function

    4.4.3   Consider a function that will provide greater benefits

    4.4.4   Determine the relative importance of the desired benefits, such as performance, quality, reliability, appearance, length of life, ease of operation and maintenance, delivery, and service

    4.4.5   Set upper and lower bounds to the desired benefits

    4.4.6   Express the resulting range of choice in a rating scale

    4.4.7   Assign weighting factors

4.5   What *should* it cost to provide the function?

    4.5.1   What is the least cost to provide the *basic* function?

    4.5.2   What is the least cost to provide the desired supporting functions?

4.6   Establish value targets

    4.6.1   Compare least cost to actual or proposed cost

    4.6.2   Identify the reasons for departures from least cost

    4.6.3   Relate the elements of worth to the corresponding elements of cost in order to identify high-value and low-value items

    4.6.4   List items or areas for value improvement

4.7   Summary of the investigation

    4.7.1   The project as it looks now

    4.7.2   The project as it should be

    4.7.3   Apply the principle of balance and proportion to the search for improvement

5.   INNOVATION PHASE I

5.1   Barriers to innovation

    5.1.1   Undue dependence on the dictates of parents, teachers, books, custom, and tradition

    5.1.2   Undue caution

    5.1.3   Lethargy

    5.1.4   Fixed ideas

    5.1.5   Unchanging attitudes

    5.1.6   Established patterns of behavior

5.2   Aids to innovation

    5.2.1   Abandon your regular patterns of thought

    5.2.2   Momentarily ignore what you learned in school

    5.2.3   Forget the way things are supposed to be done

    5.2.4   Question tradition

    5.2.5   Suspend critical judgment

    5.2.6   Seek a great number of ideas

    5.2.7   Seek a great variety of ideas

    5.2.8   Appraise the cost and risk involved in *not* taking chances

5.3   The mechanism of innovation

    5.3.1   Combination and selection

Table 10-2 *(Continued)*

    5.3.2 Imagination and forecasting

    5.3.3 Random exploring

    5.3.4 Systematic search

    5.3.5 Withdrawal

    5.3.6 Insight

5.4 Adapt to change by considering

    5.4.1 new materials

    5.4.2 new methods

    5.4.3 new customer requirements

6.   INNOVATION PHASE II

6.1 Create change by variations in the following:

    6.1.1 The means of satisfying wants; choice of approaches, intermediate objectives, choice of inputs and outputs

    6.1.2 The interaction of elements; circuit, cycle, flow network, linkages, etc.

    6.1.3 The use of time; sequence, duration, frequency, etc.

    6.1.4 The use of space; arrangement, orientation, form factors, size, etc.

    6.1.5 The disposition of matter; state, density distribution of mass, weight, etc.

    6.1.6 The use of energy; sources of power, modes of drive, etc.

6.2 Identify with the product or service—personalize the problem, "If I were a piston inside a cylinder. . . ."

6.3 Stand detached from the product or service—put the problem in perspective, "Who cares about the piston? What is it for anyway?"

6.4 Search for analogies

    6.4.1 in the life sciences

    6.4.2 in the physical sciences

    6.4.3 in our own industry

    6.4.4 in other industries

6.5 Search for new ideas, using the aids in 5.2

    6.5.1 List primitive or elementary methods

    6.5.2 Examine the absurd and paradoxical

    6.5.3 Forecast methods of the future

6.6 Examine and screen the choices developed

    6.6.1 Simulate or forecast consequences

    6.6.2 Exercise critical judgment (deferred until now)

7.   SIMPLIFY

7.1 Chart shorter and straighter paths to the following:

    7.1.1 Sense, detect, transmit, actuate, and otherwise convey information and commands

    7.1.2 Convert, transfer, and transmit energy

    7.1.3 Convert and transmit motion

Table 10-2   *(Continued)*

7.2   To accomplish 7.1 above try
    7.2.1   eliminating
    7.2.2   combining
    7.2.3   rearranging
    7.2.4   modifying
    7.2.5   substituting
    7.2.6   standardizing

8.   EVALUATION

8.1   Compare the choices developed
8.2   Select the best
8.3   Estimate gains
    8.3.1   Improvements in performance, quality, reliability, schedule, appearance
    8.3.2   Gross dollar gains
    8.3.3   Cost to implement
    8.3.4   Net dollar gains
8.4   Estimate time of payoff
    8.4.1   Time to implement
    8.4.2   Time to amortize
    8.4.3   Total payoff period
8.5   Appraise the risk
    8.5.1   Performance problems
    8.5.2   Procurement problems
    8.5.3   Manufacturing problems
    8.5.4   Market hazards
    8.5.5   Other hazards

9.   PREPARE THE PRESENTATION

9.1   Prepare reports
    9.1.1   Expected gains
    9.1.2   Expected costs
    9.1.3   Payoff period
    9.1.4   Risks involved
    9.1.5   Outline a plan for implementation
9.2   Review reports
9.3   Schedule dry runs
9.4   Revise charts
9.5   Conduct dry runs

10.   PRESENTATION TO MANAGEMENT

10.1   Welcome
10.2   Introductory remarks
10.3   Presentations by task group spokesmen

VALUE ENGINEERING IN GOVERNMENT CONTRACTS

Military procurement, in all ages, has suffered the disadvantages of an institutionalized emergency. During a war, when the issue is survival, all manner of resources are hastily converted into munitions, economy goes by the board, and waste is accepted as a necessary part of the emergency.

One bright side of the picture is that the enemy does exactly the same thing. He, too, runs the risk of indiscriminately expending resources and then running short of the one or two critical resources without which he cannot fight a war. At the end of World War II, therefore, even the winners found themselves short of resources; yet the emergency outflow valves remained wide open, though the torrent had dwindled to a trickle. To get the necessary volume in vital areas the flow had to be reduced in others. Such discrimination calls for thinking in terms of value. Instead of closing all valves by a half a turn, as happens in many a cost-reduction program, a few companies adopted a value strategy and reduced only the less productive flow in order to maintain or increase flow where it did the most good.

Another Arrow in Uncle Sam's Quiver

Instead of institutionalizing wartime design and procurement practices, the United States Government added an extra arrow to its defense quiver by using the period *between* emergencies to develop effective methods of resource utilization.

One of these methods is value engineering. It has given the United States Government a definite edge over those nations that cannot or will not take advantage of the profit motive to improve government specifications.

The government value-engineering programs reap added benefits from the national technology by rewarding those contractors whose initiative and expertise improves the *government's part* of the contractual relationship, both technically and economically.

When, as the result of a contractor's proposal and action, the government either pays less or gets better equipment for the money, the contractor is awarded a portion of the dollar gains. It is not so simple as all that, of course. The preparation, submission, and acceptance of a value-engineering change proposal (*VECP*) is complex, indeed, but the effort is well worthwhile, often better than new business, because the government pays well for improvements that can save taxpayers enormous amounts of money across the board.

How can the return be greater than that of new business? Because, thanks to his initiative and specialized skill, the contractor is allowed to improve a portion of BIG business—the biggest business there is, United States Government business. The results of his effort are multiplied by the difference in magnitude between his operation and the huge operation he is momentarily allowed to improve.

If contractor value engineering can save the taxpayers money, a vendor value program can save your company money. More important, from an industrial-engineering viewpoint, such a program can bring vendor expertise right into your house—save you a lot of dog work trying to keep up with the state-of-the-art in industry.

Commercial Applications of the Sharing Approach

The difference between value engineering by a supplier and what the supplier is paid to do anyway is that the supplier provides his own specialized knowledge to help the buyer define requirements in those areas better known to the supplier.

A well-rewarded, vendor value program can do more than save a company money. It can actually increase the technical value of purchased parts and components by having the vendor use his special knowledge and experience to look at *his* product from *your* side of the fence.

VALUE-ANALYSIS TRAINING

When the value workshop is centered around the work itself, around products that are particularly important to the company, value-analysis training is centered around the employee with a view to improving performance in all those aspects of his work that relate a product's function to its cost. The goal changes from *improve product value* to *improve employee contribution to value*. It is a teaching goal.

Consultants Can Be Helpful

If you want to start off in style, you might hire a consultant to put on a one- or two-week seminar. They bring in the additional people and equipment for a real spectacular effort, giving your training program good initial momentum.

You can, of course, put on your own condensed seminar, but your bosses haven't parted with beaucoup dollars—all at one time—to have you do it. Once the contract has been signed and the senior consultant has casually pocketed the advance check, *nobody* cancels, delays, or stands in the way of the seminar. I have used consultants to put on seminars in our headquarters areas, have sent people to public seminars, and have had an in-house training course. The section that follows arranges value-analysis material for the purpose of "changing behavior" on a day-to-day basis.

A COMPANY VALUE-ANALYSIS COURSE

Schedule

Classes meet for two-hour periods during 14 weeks. Six of the classroom sessions are followed by two-hour lab periods. In sessions 6, 7, 9, 11, 12, and 14 the combined classroom and lab work will take up a half day or a whole evening.

Timing

Running the course on company time, preferably in the forenoons, is best. When live company projects are used the results more than pay for the marked-up time of participants and instructors and there is immediate carryover into the job.

Classroom Sessions

Session 1.   Value Analysis and the Concept of Value

I. Introduction
    A. What is value analysis?
    B. Distinguishing traits of RCA value analysis
        1. Does not attempt to do the other man's work for him
        2. Simply provides a faster and more direct means of interaction among existing specialists
        3. Works under the direction of the man normally responsible for the product
        4. Contributes to "making it right the first time" and then to "keeping it right" as times change
II. How it began
    A. Substitution of scarce materials
    B. Function and structure
        1. Defining the function
        2. Evaluating the function
    C. Matters of fact and matters of value
        1. The concept of value
        2. From what is to what should be
III. Definitions
IV. Aspects of value
    A. Use value
    B. Esteem value
    C. Market value
    D. Exchange value
V. Components of exchange value
    A. Customers with wants and resources—a potential market
    B. Utility to such customers—the product must be worth something to them
    C. Scarcity or difficulty in attainment—the product must be hard to get
    D. Total cost to the customer—an inverse component of value. Given difficulty in attainment, the customer wants to pay the least to overcome that difficulty

VI. How product value was rent asunder
    A. One-man craftsmanship
    B. The modern factory
        1. Specialization
        2. Departmental separation
        3. Lost communication
        4. Stale information

Session 2. The Multidiscipline Approach

  I. Reintegrating product value
    A. Coordinating the work of specialists
        1. Function of a committee
        2. Function of a team
    B. The value task group: a multidiscipline team
        1. Combination of pertinent skills
        2. Orchestration of effort
        3. Instant communication
        4. Joint updating on the state of the art
    C. How the task group works
        1. Methodical search for product information
        2. Analysis of customer wants and what the product should *do* to satisfy them
        3. Relating benefits to their cost
        4. Deliberate search for new ideas
        5. Evaluation, comparison, and selection of options
        6. Detailed plan of implementation
  II. Value analysis and product worth
    A. Demand for better products
        1. Schumpeter's warning
        2. Value of a free society
        3. Needs or desires
    B. Meeting the demand for better products
        1. Model-T to Lincoln Continental
        2. Volkswagen's Bare Bug, Love Bug, and 1500 Fastback
        3. Rolls Royce's use of value analysis to improve the best
 III. Value analysis and product cost
    A. Good costs and bad costs
    B. Nature of costs
        1. Time
        2. Space
        3. Matter
        4. Energy
        5. Human effort
    C. Immediate importance of cost reduction
    D. Long-range benefits of product improvement

    E. How a value-analysis team does product improvement *and* cost re-reduction in relation to each other
        1. Representing both technical and economic skills
        2. Working *for* the man responsible and not around him
        3. Offering information and service but not pressure or direction
    F. Upstream, downstream, or both
        1. Providing timely, marketing procurement, and manufacturing information to the new product designer
        2. Updating yesterday's perfect products to stay ahead of the competition
IV. Management principles of value analysis
V. Body of knowledge underlying the practice of value analysis
VI. Working in harmony with company goals and objectives
    A. Company goals
        1. Sales
        2. Profit
        3. Product leadership
        4. Market penetration
        5. Community service
    B. Company objectives
        1. Survival
        2. Return on investment
        3. Growth
        4. A good public image
VII. Goals and objectives of value analysis
    A. General objective
        1. To improve product value
    B. Specific goals of value analysis
        1. Improving the use of time
        2. Improving the value of information
        3. Increasing the effectiveness of communications
        4. Forestalling people problems
    C. Specific objectives of value analysis
        1. In what we buy the objective is to improve the relationship between the total cost of a purchased item and what it is worth to us
        2. In what we make the objective is to increase the margin between cost to manufacture and sale price
        3. In what we sell the objective of value analysis is the same as that of everybody else in the working force, to increase the gross margin between total cost to us and total revenue

Session 3. Your Own Value Analysis

I. We all do it intuitively
    A. Wives, when they shop for the house

      B. Everyman, when he makes an important purchase
      C. A student, when selecting courses
II. Who can profit from formal value analysis?
      A. The professional buyer on his way up
      B. The design engineer
      C. The manager
      D. Value analysis in the factory

Note. Part I of this session is meant to provide participants with the tools of value analysis in solving their personal problems of choice, such as investments, college for the children, where to go on vacation, and so forth. Once developed this habit has direct carry-over into the job, particularly in scheduling their own work. Part II offers an overview of value analysis throughout the plant.

Session 4.   Value Analysis in Your Operation

I. This session covers, in detail, the application of value analysis to the trainee's particular discipline
      A. Finance
      B. New product planning
      C. Marketing
      D. Design engineering
      E. Purchasing
      F. Manufacturing
      G. Packaging and shipping

Note. This is the specialized portion of the course. Session 3 provided a brief description of how value analysis is practiced across the organization.

This session offers a detailed description of how it is done in the pertinent one of the disciplines listed above. A course meant for engineers would include only Item I-D at this point. If members of Marketing and Finance were taking the course, Items I-C and I-A would be presented for their benefit in separate classrooms.

In a plant-wide course, seven different individual versions of Session 4 would have to be presented.

Session 5.   How It Works

I. Requirements
      A. Acquiring the necessary skills
      B. Face-to-face communication
      C. Touching base with the specialists
II. The value task group
      A. Selection of projects
         1. Jobs that must advance the state of the art
         2. Jobs that must be delivered ahead of schedule
         3. Jobs that cost more than they should

    B. Different groups for different tasks
       1. Search for new materials and products to buy
       2. Search for new products to make
       3. A new product support group
       4. A design support group
       5. A product updating group
    C. Selection of participants
       1. Achieving a balanced combination of pertinent skills
       2. Verifying feasibility of dependable attendance
    D. The task group in action
       1. Riddle of group dynamics
       2. Management principles applied at this stage
          (a) direct motivation
          (b) interacting skills
          (c) objectivity
          (d) effective communication
III. The value analysis job plan
    A. Preparation
    B. Information phase
       1. Principles
          (a) *adequate* information
          (b) selectivity
       2. Task
          (a) investigation
    C. Analytic phase
       1. Principles
          (a) direction
          (b) responsibility
          (c) limited resources
          (d) economy
       2. Task
          (a) analyze the product's function
    D. Creative phase
       1. Principle
          (a) balance and proportion
       2. Tasks
          (a) innovation
          (b) simplification
    E. Evaluation
       1. Verification
       2. Comparison
       3. Choice
    F. Presentation
       1. Report
       2. Visual aids
       3. Making the presentation

G. Implementation
   1. Providing enough information
   2. Forecasting consequences
   3. Obtaining
     (a) authorization
     (b) skills
     (c) facilities
     (d) resources
   4. Scheduling
   5. Follow-up
   6. Reporting results

Session 6. The Value of Information

I. To be information it must be as follows:
  A. New
  B. True
  C. Useful
  ... Otherwise, it is noise

II. Definition of information terms
  A. Vigilance
  B. Alertness
  C. Raw information
  D. Intelligence (G-2)
  E. Accuracy
  F. Precision

III. Information phase of the value job plan
  A. Appraise tasks and resources
  B. Appraise skills, man-hours, and information available
  C. Appraise task group projects
    1. Identification—what is it? Part, model, or contract number and description
    2. A quick look at the function—what does it do?
  D. What does it cost to provide the function?
    1. Cost per unit and quantities involved
    2. Cost breakdown—identify the elements of cost
  E. Characteristics of good information
    1. Novelty
    2. Truth
    3. Freshness
    4. Predictive potential
    5. Selective power
    6. Good signal-to-noise ratio
    7. Identifying marks
  F. The search situation
    1. Data needed for action

2. Resources for the search
3. Scope of the search
4. Assigning search sectors
5. Sequencing search sectors
6. Finding suitable search methods
7. Control to avoid repetition

IV. Turning information into gold
    A. Verifying requirements and constraints
    B. Validating tolerances and allowances
    C. Updating and looking ahead
    D. Searching for the following:
       1. Technical break-throughs
       2. New materials
       3. New methods
       4. Changes in customer needs
       5. New markets
    E. Identifying additional information required

V. The principle of selectivity
    A. Selective potential of information
    B. Other applications of the principle
       1. Selectivity in the choice of tasks
       2. Selectivity in the choice of skills

Session 7. Analytic Phase of the Job Plan

I. Analysis of the project for concentration of effort
    A. Apply the principle of selectivity in directing effort to the most promising areas
    B. Plan the expenditure of time and effort in relation to workshop skills, supporting group skills, vendor participation

II. What does it do? Define the function
    A. Pinpoint the function's most important aspect, using only a verb and a noun
    B. Ask "why" to raise definitions to higher levels of abstraction
    C. Ask "how" to reduce definitions to lower levels of abstraction
    D. Consider one function at a time
    E. Search for measurable properties
    F. Define the function at the highest level of abstraction within the scope of the company's business

III. Primary, secondary, and equally important functions
    A. Determine the interdependence of functions
    B. Identify independent functions

IV. Evaluate the function. What is it worth to the customer?
    A. Determine the relative importance of the elements of worth (also called *benefits*)
       1. Performance

      2. Safety
      3. Quality
      4. Reliability
      5. Appearance
      6. Length of life
      7. Ease of operation and maintenance
      8. Delivery
      9. Service
  B. The concept of utility
  C. Set upper and lower bounds to the desired benefits
  D. Express the resulting range of choice in commensurable units
  E. Assign measures of relative importance to the desired benefits—weighting factors

V. What *should* it do? Identify opportunities for improvement
VI. What *should* it cost to provide the function?
  A. What is the least cost to provide the basic functions?
  B. What is the least cost to provide the desired supporting functions?
VII. Establish value targets
  A. Compare least cost to actual or proposed cost
  B. Identify the reasons for departures from least cost
  C. Relate the elements of worth to the corresponding elements of cost in order to discover high-value and low-value items
  D. List the most promising items or areas for value improvement
VIII. Summary of the investigation
  A. The project as it looks now
  B. The project as it should be
  C. Apply the principle of balance and proportion to the search for improvement

Session 8. Theory of Creativity

I. Inventive traits (latent in all of us)
  A. Receptivity
  B. Daring
  C. Discernment
  D. Curiosity
  E. Convergent thinking
  F. Divergent thinking
  G. Fluency
  H. Flexibility
  I. Tolerance
II. Environment for invention and discovery
  A. Stimulating environments
    1. Intellectual freedom
    2. Economic safety
    3. Tolerance of contrary views

     4. Humor without ridicule

     5. Animated discussion without hostility

III. Process of invention and discovery

    A. The classic steps

       1. Preparation

       2. Withdrawal

       3. Insight

       4. Verification

    B. The controversy on "brainstorming"

       1. Individual creativity

       2. Creative groups

Session 9.   The Practice of Innovation

  I. How to welcome new ideas

    A. Criticism postponed

    B. Criticism transformed

 II. Freeing your inventive personality

    A. Barriers to innovation

    B. Aids to innovation

    C. The mechanism of innovation

III. The task of innovation

    A. Adapting to change by considering the following:

       1. New requirements

       2. New materials

       3. New methods

    B. Creating change through variation in the following:

       1. The means of satisfying wants

       2. The interaction of elements

       3. The use of time

       4. The use of space

       5. The disposition of matter

       6. The use of energy

IV. The task of simplification

    A. Trend toward greater complexity

    B. Identifying useful complexity

    C. Removing excess complexity

    D. How order tames complexity

    E. Economy of action

       1. Finding the most direct method for sensing information

       2. Finding the most direct method for converting energy

       3. Charting more direct paths for conveying information, transferring energy, and transmitting motion

    F. Specific techniques of simplification

       1. Combination

       2. Rearrangement

3. Substitution
4. Elimination
5. Standardization

Session 10. Evaluation

I. Numbering, measuring, weighing
   A. Quantity and number
   B. Definitions of measurement
   C. Function of measurement
      1. Conserve resources
      2. Contribute repeatability
      3. Quantify benefits
      4. Quantify cost
      5. Monitor progress
II. Scales of measurement
   A. Nominal
   B. Ordinal
   C. Interval
   D. Ratio
   E. The bounded interval scale of value analysis
III. The COMBINEX Aid to Choice and Decision
   A. Identify the benefits
   B. How to compare apples and oranges
   C. Quantify the benefits
      1. Effectiveness and utility
      2. Efficiency and value
   D. Setting upper and lower bounds
   E. The additivity assumption
      1. Independent benefits
      2. Interacting benefits
      3. Excluding the inadequate
      4. Excluding the excessive
   F. Weighting techniques
   G. The COMBINEX scoreboard
      1. Effectiveness of each choice
      2. Contribution to each weighted benefit
      3. Relative value of the choices
IV. Appraising value-analysis options
   A. Benefits and advantages
   B. The three baskets
      1. Time
      2. Money
      3. Risk
   C. An economic definition of risk

Session 11.   Business Comparison

I. Expected gains
    A.   Product improvement
    B.   Schedule improvement
    C.   Gross dollar savings

II. Investment
    A.   Skills
    B.   Time
    C.   Dollar cost of implementation
        1. Loss of product during changeover
        2. Scrap or rework
        3. New tools
        4. More costly materials
        5. Engineering cost to implement
        6. Manufacturing cost to implement

III. Net gains
IV. Return on investment
V. Payoff period
VI. Break-even point
VII. Risk
    A.   Performance problems
    B.   Procurement problems
    C.   Manufacturing problems
    D.   Market hazards
    E.   Other hazards

Session 12.   Reporting Results

I. Prepare the report in the language of those who will read it
    A.   Who will authorize the action?
    B.   Who will implement it?
    C.   Who should see the draft?
    D.   Who will verify the figures?

II. Sections of the report
    A.   The project
    B.   Problems and opportunities
    C.   Proposed action
    D.   Cost, risk, and net gains
    E.   Return on the effort
    F.   Comparison with other investment opportunities
    G.   Business-oriented summary
    H.   Plan of implementation

III. Review the report
    A.   With team members
    B.   With workshop leader

      C. With Finance

      D. With the man normally responsible for the product

 IV. Prepare the presentation to management

      A. Use key points of report

          1. The team

          2. The task

          3. Problems or opportunities

          4. Recommendations

          5. Analysis

             (a) what will this buy us

             (b) what will it cost us

             (c) net gains

             (d) risk

             (e) pay-off period

             (f) break-even point

          6. Highlights from plan of implementation

  V. Prepare the visual aids

 VI. Prepare a detailed plan of implementation

 VII. Conduct dry runs

VIII. Make required revisions

Session 13. *Communicating Information: The Verbal Presentation*

  I. Requirements for communication: the transmitter

      A. Use the full capability of your voice

      B. Avoid strain

      C. More than audio—spirit!

      D. Spare the corn

      E. Importance of posture

 II. The receivers

      A. An audience listens harder than an individual

      B. Talk to *them,* not to the blackboard, the flipcharts, or your notes

      C. Do not read what is in the charts or slides—the audience can read and write

      D. Management wants bottom-line answers, not details

III. The message

      A. Information must be new

          1. Repeat only to summarize

          2. Repetition is a form of marital pressure called nagging

          3. To emphasize a point pause, slowly roll out your words; or holler; or jump; BUT DO NOT REPEAT!

      B. Information must be true

          1. Give source of your facts

          2. Acknowledge what you don't know

      C. Information must be useful

          1. Details, not immediately useful to the audience, are only noise

          2. Use humor only when it conveys real information

IV. Summary
- A. Volume
    1. Use enough air
    2. Talk from the chest
    3. Let the words roll out loud and clear
- B. Posture
    1. Face the audience squarely
    2. Stand up straight
    3. Hold your head high
    4. Don't hide behind or embrace the rostrum—better, no rostrum
- C. Direction
    1. Talk to the audience
    2. Do not talk to the blackboard or the charts
    3. Do not talk to your notes
    4. Read, then face them and tell them
- D. Conciseness
    1. Throw away the oyster shells
    2. Eat the oyster
    3. Give your audience the pearl

Session 14. *Implementation*

I. A good proposal should be easy to select
- A. Helping management choose what to implement
- B. Providing enough information
- C. Making an honest forecast of the consequences
- D. Finding the right sponsors

II. Authorization

III. Skills and facilities

IV. Additional resources
- A. Marketing money
- B. The chicken and the egg money problem
- C. Direct labor funds
- D. Procurement funds
- E. Finance, where the funds are

V. Plan of implementation
- A. What must be done
- B. Who will do it
- C. Where will it be done
- D. How will it be charged
- E. When will it be done
    1. Starting date
    2. Reporting periods
    3. Completion date
    4. Gantt chart for implementation

VI. Engineering work order

VII. Bill of materials

      1. Raw materials
      2. Purchased parts
      3. Tools

VIII. Manufacturing work order

  IX. Simple reporting forms

   X. Follow through

      A. Name of person responsible for monitoring program in each department

      B. Hours per week he is to spend on it

      C. Reporting dates to his superior

      D. Reporting dates on which department managers will report progress of implementation to general manager

  XI. Reporting final results

Lab Periods

Each of these periods follows the session of the same number.

   6-L   Information phase
   7-L   Analytic phase
   9-L   Creative phase
  11-L   Comparison and choice of recommendations
  12-L   Preparation of project reports and visual aids
  14-L   Team presentations

XI

RECOGNIZING OTHER
IMPROVEMENT ACTIVITIES

It would be nice to say that the interaction between the various improvement activities is similar to the mutually beneficial and harmonious relationship between mosses and lichens or ants and aphids.

Many years ago I felt that this was really the case, but it just happens that I was lucky. Working for a small, tightly knit industrial plant, I did not realize that the improvement activities were all there in their nonspecialized form.

Later, as a systems engineer for a larger company, I was lucky again; no conflict with other improvement functions; nobody tried to help me optimize my projects.

Then the first company invited me back as chief mechanical engineer. With departmental responsibility came department-manager authority. The company had grown and now had a number of staff specialists, but conflict between staff improvement functions and a line manager is only a minor irritation to the latter, though a dangerous morass for the staff specialist. As manager of manufacturing in another small company I was still boss of my own operation.

While I was reflecting on this pleasant aspect of line management, fate gave me a new job as a staff specialist. I became an explorer of morasses, navigator in crowded harbors, and gold miner at large. I headed value engineering in one of the divisions of a large corporation.

Exploring the morass of staff conflict with line management is simple enough; you find solid ground through the swamp. This solid ground is called *service,* as distinguished from help and direction. There simply should be no conflict between line functions and supporting functions, between the man responsible and the services available to him for doing a better job. To work around this morass useful information and sound motivation must replace pressure.

Navigating a crowded harbor is another story. Here is where the various improvement efforts could collide, but we can learn much about avoiding collisions from the successful small plant.

DEVELOPMENT OF THE IMPROVEMENT FUNCTIONS

Many of us look back longingly to the Arcadia of the small plant in which pride of craftsmanship ensured quality, industriousness improved production, and ingenuity created designs of merit at reasonable cost.

Obviously there is no need for improvement effort in such an organization, just as there was no need for improvement effort in the Arcadia to which the Greeks, after they had been clobbered by the Doric invasion, looked back longingly. Dreamland Arcadia, with its advanced culture and its secure cities "rich with gold," must have needed some sort of improvement or the fierce but crude Dorians could not have conquered it. Archaean Greece had grown fat and sleek. It had provided a good life to remember, but it could not meet competition.

The same is true of the old-fashioned plant of fond memory. Where is it today? Either it adopted modern improvement methods and became a successful modern plant or it was taken over by one that had. Exceptions are the small plants that failed altogether and the few true Arcadias, in which outstanding management and an outstanding working force can turn out a consistently good product. Industry, by and large, cannot depend on outstanding individuals. If there were enough of them to go around, they wouldn't stand out. What industry needs are methods to make people, as they are, turn out consistently good products.

The question is not whether we need the various improvement activities, but how can we make them more effective, considering that each can work, with some measure of success, in the specialized field of the others.

There are three dangers to guard against:

1. Reduced value of the output, as when a miner neglects his own diamond mine to encroach on the neighbor's gold mine.
2. Waste of resources by duplication of effort, as when two young men

spend the evening fighting over one girl, neglecting several equally desirable girls.

3. The least logical but most damaging of all—arousing the animal aggressiveness in people by invading their organizational or professional territory.

This last does not mean simple rivalry between specialties. Such rivalry can be handled by the rules of sportsmanship. The danger here is actual invasion—the husband pushing his wife aside to bend over the cookstove and put more salt in the soup, the engineer telling the buyer how to buy, or the buyer telling the engineer how to design.

To cope with these dangerous situations we can learn something from the small plant, both from the idealized small plant of fond memory and from the successful small plants that have made it and made it without organizationally defined improvement activities.

There is no question that the improvement skills are there. If they weren't, the small plant would not be successful.

There is no question, either, of the need for specialization. In the small plant I mentioned in the beginning—my own pet Arcadia—we had industrial engineers, systems engineers, and a company president who understood and used value improvement and operations-research techniques; but there were no departments of industrial engineering, systems engineering, value improvement, and operations research. The skills were there, but they were not organizationally segregated, and no field of activity—no territory—was assigned to each improvement skill.

A look into the small plant that succeeds without formal improvement activities reveals the following:

1. Skills directed toward improving the decision process, improving management methods, improving manufacturing methods, and improving the value of the product or service.

2. No organizational identity or territorial preserves are given the improvement activities.

3. Little or no overlapping in the improvement effort.

Now, what can we learn from this situation? We already know that the situation is rare, that there are few successful plants today without formal improvement operations. Why? Obviously because there is no industrial survival without change for the better, not in a rapidly advancing technology. Why formal improvement activities? Because lack of organizational identity means lack of performance measurement and cost visibility. There is no mechanism for relating the accomplishments of the improvement effort to its cost.

Not knowing who is doing what is hardly an advantage, but harmony in the improvement effort *is*. Will identity and control of the improvement functions necessarily impair this harmony?

Not if we leave out the territorial concept. Smokey the Bear, the fire-fighting conservationist of the Forest Rangers, is a good example of the chief of an improvement function who has no territorial pretentions. Neither Smokey the Bear nor the Forest Rangers protest when the Girl Scouts and the Camper's Association get into the conservation act. Preventing forest fires and getting others to prevent them is not their territorial preserve, it is their overall goal and they welcome anyone who will help them attain it.

The principal lesson we can learn from the small plant is that the lack of territorial preserves among the improvement functions can promote harmony. This lesson leads us to look at the other side of the picture.

Table 11-1   Characteristics of the Struggle for Territory

1. Defending area of influence against encroachment
2. Defending existing knowledge from the impact of outside ideas
3. Defending occupational jurisdiction
4. Extending area of influence
5. Extending existing knowledge
6. Extending occupational jurisdiction

SECTARIANISM AMONG INDUSTRIAL DISCIPLINES —A REAL PROBLEM

In order to understand the full meaning of sectarianism we must take a quick look at its history. Among the many blessings bestowed on us by Imperial Rome, such as Roman law and Roman roads, we inherited two dangerous concepts: one, that to be safe a nation must ultimately rule the whole world; and, two, that for a religion to be right all other religions must be wrong. The first made it possible for the Romans to rule nearly everybody they knew, at the cost of becoming slaves of their own tax collectors; and the second brought about endless religious wars.

We cannot blame sectarianism on the Golden Rule or the Sermon on the Mount. Western sectarianism was the natural result of the Roman drive for cultural domination and for empire—and this brings us right back to the territorial imperative (Ardrey) and to animal aggressiveness (Lorenz).

If we can divest the various improvement functions of tendencies toward cultism and territorial pretensions, we can achieve the harmony existing in small plants without foregoing the advantages of organizational identity and administrative controls.

SYSTEMS ENGINEERING
—A LINE IMPROVEMENT FUNCTION

A problem in mathematical thinking and in the theory of measurement hinges around the distinction between *more* and *better;* that is, between magnitude, extent, or intensity, on the one hand, and degree of desirability, utility, or effectiveness, on the other.

We carelessly say, "The highest possible quality for the least possible cost gives us the greatest value." Here we are talking about magnitudes, and incompatible magnitudes at that. The *highest* possible quality may be excessive quality—more than anybody could want or use—and high quality is seldom compatible with the *least* possible cost—which is no cost at all.

We then explain that we mean the highest possible quality *for the cost.* When asked "for *what* cost," we mumble something about the cost of good quality. Now good quality is a far cry from highest possible quality, but we have gained one advantage. Like it or not, we have been forced *to fix* one of the two variables—to make it hold still while we work with the other.

We are on sound ground mathematically when we hold one variable still while we maximize or minimize the other. Maxima and minima, however, relate to *most* and *least* and—with the possible exceptions of the state of Texas and of flea circuses—bigness or smallness may not be advantageous in themselves. Given a numerous selection of girls for an evening date, neither the biggest nor the smallest girl may be the most desirable.

The question usually is not one of *the most* or *the least* but of the best for a given purpose, not one of brute effort toward the maximum or the minimum but of a more subtle weighing and comparing in order to achieve the optimum—*the best* relationship among the variables.

Optimization—approaching the best—is a synonym for improvement, and the improvement function specifically entrusted with the task of optimization is systems engineering. Optimization here involves selecting the right components or subsystems and arranging them in the best relationship in order to make up a successful system. Such systems, which range from small cameras and their interchangeable lenses to satellite communications systems, constitute the grist of the systems engineering mill, a mill that grinds the grain it is given to grind and does not go around trying to grind everything else.

Because it works for itself, so to speak, making decisions that primarily affect its own recognized sphere of activity, systems engineering seldom creates problems arising from territorial invasion—or worse—from the missionary infiltration that is implicit in trying to improve the activities of others.

In specifying the components it wants, systems engineering does occasionally participate in optimization tasks that are entrusted primarily to component designers, yet its recommendations seldom encounter hostility. Why? . . .

The true professional, whose qualifications are recognized and who commands the support of a well-organized and well-verified body of knowledge, does not need to urge, exhort, or preach—he simply demonstrates.

The unconsecrated improvement specialists in the small plant are professionals to begin with—industrial engineers for the most part—or highly skilled and respected craftsmen. In either case their improvement efforts are based on a sound understanding of the particular improvement task that is usually part of their regular work.

Because of their many chores, these men must move along the improvement path so fast that they cannot linger over the pitfalls long enough to fall in. There is no time to preach a sermon or to criticize rival cults, no time to claim territorial jurisdiction. If someone is plowing one end of their field, they thank God that it is being plowed at all and rush to plow the other end.

This wholesome situation, which prevails when people are really busy, is another of the small plant's advantages—*the understaffed improvement function seldom gets in anybody's hair.*

This is not to advocate inadequate manning, but, since there is usually some discrepancy between a need and the manpower resources assigned to meet that need, we should take care that the discrepancy is on the side of leanness rather than excess.

I will now review the pitfalls and aids we have identified so far because we need to understand them in order to arrive at the natural distribution of effort among improvement functions.

PITFALLS

Cultism is the worst pitfall because it breeds lasting hostility toward the improvement function and toward its proponents. Based on the assumption that they have a divine mandate from The Man Upstairs, the cultists describe their improvement function as "the best thing that came down the pike." It can grow hair on billiard balls, optimize the parking lot, and redirect the space program toward a more profitable galaxy. This can be taken with a grain of salt, but when the cultist implies that his victims—the line people—cannot distinguish between the company's goals and a hole in the ground, that they have to be made cost conscious and told that profits are desirable, he is on the road to martyrdom and may well end up in the belly of a lion.

Territorial pretentions constitute a pitfall characterized by the phrase, "Your function belongs in our department," or the warning, "Don't you dare set foot in our area."

One of the most gratifying experiences I have had in my career took place when I called up the manager of the management engineering function to determine the relationship between my value-analysis program and a fine work-simplification program that came under him. His first words were, "Let's get this straight; *we don't intend to take over your function.* We're so busy we can use all the help we can get. If we take on a job and it looks like a job for value analysis, we'll throw the ball to you."

Since then we have been throwing the ball to each other like a well-coached basketball team. I had a similar experience with the vice president who heads our management information systems. Is all this simply because we are very fine people?

Of course!

But, in addition to that, are there other reasons? . . .

TWO AIDS TO HARMONY IN THE PLANT

Professionalism is more than an aid, it is the basis of a sound improvement function. Line managers do not resent professional service backed by solid know-how. They do resent exhortation and pressure which all too often implies that they and their people are inefficient, indecisive, not cost conscious, and ill informed.

Leanness

Peter Drucker identifies overstaffing as one of the time-wasters in industry, pointing out that overstaffing ties up the time of senior executives with people problems—too many people without enough to do.

A lean improvement function which replaces cultism with solid know-how and territorial pretentions with service has an excellent chance of success, *provided it works on the right task.*

Before we tackle *that* requirement, let us look at some of the improvement functions.

IMPROVEMENT FUNCTIONS
—THE WHOLE ARRAY

Design engineering and research and development are primarily innovative functions. They have the very difficult task of going from nothing to something. They *create* what others improve. The easier task of going from something to something better can be broken down into a great variety of functions, depending on the usefulness and feasibility of specialization.

When a company has the need and the resources to staff a full array of improvement functions, the list could run something like this:

> Industrial engineering
> Management engineering
> Systems engineering
> Operations research
> Value analysis
> Management information systems

. . . and other functions, such as manufacturing engineering, quality control, configuration control, work simplification, human-factors engineering, reliability engineering, design review, short-interval scheduling, systems and procedures, methods improvement, cost-reduction suggestion program, zero-defects program, standardization, overhead control, inventory control, and more to come.

Such an array obviously calls for skillful navigation in a crowded harbor, but it also calls for more than that. It calls for good planning of harbor traffic and for well-enforced harbor rules. Each function must know where it is going and must obey the Rules of the Road so that all can get to their destinations without colliding.

DISTRIBUTION OF TASKS
AMONG IMPROVEMENT FUNCTIONS

The traditional improvement profession in industry has been industrial engineering. Its natural task is to improve the efficiency of industrial production by devising better ways to make a product or render a service. Since all companies use internal services, that aspect of industrial engineering which pays particular attention to services has been applied to improving management methods and to advancing the science of management itself. In this sense the function is called management engineering. This gives us two clearly delineated improvement functions with their tasks.

Industrial engineering develops better ways of making products or rendering services.

Management engineering develops better methods of management and administration.

Systems Engineering

Although a ship, a locomotive, and an oil refinery constitute engineering systems, the term systems engineering came into being with the development of complex military weapons systems. Its activities have been described earlier. In general, systems engineering optimizes the performance of physical systems.

Operations Research

The scientific analysis of large-scale military operations in World War II led to the use, in the British Isles, of multidiscipline scientific teams. These teams of "operations analysts" in turn developed techniques of inquiry, measurement, and comparison which proved valuable in planning military operations. The procedure, now more frequently called *operations research*, proved useful to the United States Armed Forces and to industry in general. Operations research provides quantitative information for management decisions and offers techniques for comparing choices, forecasting probable outcomes, and aiding in the selection of fruitful courses of action.

As both business and technical people in industry began depending more and more on timely information, the volume of such information outstripped the traditional methods of exchanging, comparing, and evaluating it. This has led to the harnessing of computers as day-to-day tools in management and to the developing of management information systems which, of necessity, encompass all activities of a corporation.

Management information systems tie together both technical and business activities into an integrated information network to provide management with rapid and reliable tools for planning and decision making. In the case of on-line, real-time computer systems it makes possible the immediate implementation of decisions.

RULES OF THE ROAD—A SUMMARY

INDUSTRIAL ENGINEERING.   Works on improving the methods for making a product or rendering a service.

MANAGEMENT ENGINEERING.   Works on improving management and administrative methods.

SYSTEMS ENGINEERING.   Optimizes physical systems.

OPERATIONS RESEARCH.   Provides business planning models and technical models to aid in choosing courses of action.

VALUE ANALYSIS.   Optimizes product value.

MANAGEMENT INFORMATION SYSTEMS. Integrates the information, decision making, and control functions into a system that automates routine tasks and releases the time of decision makers for the better exercise of judgment.

INTERFACE WITH THE OTHER
IMPROVEMENT DISCIPLINES

There are some sportsmen who would not play football unless they threw every pass, received every pass, and made every touchdown—a trait more suitable for the prize ring than for the industrial team. As part of the industrial team value analysis must not only do what it can do best, it must let others do what *they* can do best.

The Natural Field of Activity

Diverse activities can be carried out in perfect harmony in the same department and in the same building, provided *what they do* does not conflict.

Fields someone else is plowing are Engineering and Manufacturing. These areas are fully occupied and adequately manned; invading them implies duplication of effort and yields only leavings. Many worthwhile improvement activities are properly and safely ensconced within these fields: reliability and configuration control in Engineering, work simplification and methods improvement in Manufacturing, to name only a few.

The Rich, Fallow Land. The best pay dirt for value analysis is not in these well-plowed fields but in the rich soil left fallow when the components of value were separated from one another and assigned to diverse specialists.

Today specialists on utility often have 100% technical competence in their fields. Specialists on cost are fully competent in the difficult and complex area of cost reduction. The trouble is that, having 100% technical competence on utility and 100% technical competence on cost, many a company still injects 50% guesswork. Here are some reasons:

1. Specialists on utility and on cost work independently of each other, guessing the effect of their acts on each other's specialty.
2. Neither one nor the other realizes that the *relationship* of utility to cost is what constitutes value, not the isolated magnitude of either.
3. The company has no trained value specialists whose area of operation *is* that relationship.
4. Product improvement and cost reduction are a tug of war instead of a joint effort.

"But," says the aggrieved department head, "this is the way we have always done it."

Alas! It is! The way we have always done it may have been good enough for the first half of this century, but in the third quarter, which is ending now, this loose approach is beginning to fail. It is too slow and too costly for competition with nations whose "good old-fashioned economy" was wiped out in World War II, who started from scratch, leaving the old ways behind.

We do not have to abandon the old ways, good and bad alike, but we *do* have to examine them. Comfortable and pleasant they may be, but can they cut the mustard today? Can they compete with the new ways, made possible by freedom from the past, in the new industrial nations?

Finding answers to these questions is one of the tasks of improvement disciplines. The answers themselves are not too hard to find, but voicing them takes courage, tact, and persuasiveness.

Even more courage is required to *act* on these answers—to abandon snug cocoons and venture into the world, boldly spreading wings, instead of peeking out of the cocoons like timid pupae unwilling to leave their cozy homes. The improvement disciplines must be orchestrated in such a way that they do not all rush in at once, crushing cocoons to the right and left. The right discipline must be aimed at the right cocoon.

It must be told to unravel it delicately and to demonstrate the advantages of turning from a larva into a butterfly; the advantages of adaptation and mobility.

How does an improvement discipline make such a demonstration?

By unfurling its own wings and flying a little bit, by abandoning its own set ways, experimenting, innovating, and analyzing its own performance; *then* coming up with something really useful to others.

Appendix A

MATHEMATICS OF RESOURCE USE

When Galileo taught physics, as professor of mathematics at Padua, these sciences were considered intellectual exercises, rather than tools— except by Galileo! He made telescopes, pinned down laws of nature, discovered Jupiter's satellites, and found mountains on the moon. But he was jailed by the Holy Office of the Inquisition.

CLASSICAL OR MODERN MATHEMATICS?

The example of Galileo in jail sent scholastic mathematics back into its shell so that even today classical mathematics is sometimes accused of growing up in an ivory tower. However, there is really nothing wrong with that.

After all, beautiful princesses, innocent, gentle, and untainted by greed, grow up in ivory towers. Freed from the limitations of specialized application, classical mathematics was able to develop into the universal tool that it is today, thanks, in part, to the ivory tower which made true abstraction possible.

In this ivory tower mathematics has grown up as a *princess* and not as the handmaiden of science. At Bologna, the first European university, the physical sciences were taught in the Faculty of Mathematics as they were later taught at Paris.

Sometime during the twelfth century France cut down on foreign aid and many English students at the University of Paris had to go home.

They gathered at Oxford and formed a corporation to continue their studies. Part of the cultural loss that must accompany such a change-over was the demoting of mathematics from princess to handmaid.

The tendency among nonmathematicians, who suddenly discover the value of mathematics in their own specialty, is to demand that the classical curriculum be curtailed to make room for ad hoc combinations of special mathematical techniques identified by decidedly unclassical names.

The suggestion is that the princess be evacuated by helicopter and that the ivory tower be demolished so that these creatures can enjoy a cozy game of dice in the ruins. This argument is offered in the name of progress, and it is resisted in the name of rigor; but arguing about the relative merits of "modern" versus "classical" mathematics is as misleading as it is futile. *Versus* is the wrong operator; it postulates conflict. Replace it with a *plus* (+) and we are in business.

The princess, too, can forget *versus*, send for the lusty yeomen from the royal forest, and invite them up to the ivory tower. They are linear programming, the Monte Carlo method, game theory, queuing theory, graph theory, and the theory of relations. Admittedly it is too late to have them grow up pure, but we can give them a good bath. The youths have noble traits of character worthy of the princess, but the bath *is* necessary to remove the rampant empiricism concealing their virtues.

You will find that, once cleaned up, these lusty yeomen will turn out to be well-born children of the nobility, stolen in their youth by industrial pirates or military robber barons, and taught such vile tricks as catching foxes without hounds, catching trout with nets, and putting electric starters on cross-country motorcycles.

Let's look at them first in their yeoman's garb. The linear programming of today stems from W. W. Leontieff's input-output method of economic analysis. This had replaced the French technique for solving scarcity problems with simultaneous equations. It was not until George B. Dantzig, of the RAND Corporation, developed a practical algorithm for working out the computation that linear programming became an accepted tool in military and industrial planning.

As you see, under the yeoman's garb of a militarized economist, we have a perfectly legitimate, well-born child of linear algebra and matrix theory.

OPERATIONS RESEARCH:
THE MAGIC OF NUMBERS

Operations research provides the magical utensils for converting vague hopes into feasible objectives, hidden fears into tangible foes to be conquered, and uncertainty into mathematical odds. By carefully combining the various disciplines that can be brought to bear on a problem operations research can be used to apply human skills as a collective effort instead of piecemeal. Here is an example.

In World War II, at the lowest ebb in the Battle of Britain, the allies were hopefully saying to themselves, "We will win this war yet. Wait until our economic power produces the preponderance of weapons we need."

Stark reality brought them up sharp. There was no time. The only answer was to use the *existing* weapons and equipment more effectively.

Face to face with the truth, the allied high command decided that an objective appraisal of the situation had to be made at once. The top brass knew that they themselves were committed to their special tasks far beyond any hope of overall objectivity. Scientists, being presumably intelligent and objective, were hastily gathered into teams to analyze the various aspects of the U-Boat blockade and the air blitz. The right kind of scientists—specialists in submarine and air warfare—were busy working on submarines and airplanes. The wrong kind of scientists would be better than no scientists at all. And they were. Much better! Not being emotionally involved with the problem, they asked the right kind of questions.

"What do submarines do?" asked an anatomist.

The brass was appalled. "You mean," someone demanded, "you don't *know*? . . . !"

"Not what I *should* know, Old Boy." The answer came with the imperturbable equanimity of a man who looks at life across a dissecting table. "I take it these submarines launch torpedoes," the anatomist began dissecting the problem. "How many do they carry? How many submarines are on station? How long can they stay under? How many ships in the convoy? How fast? How far? Man, we need *numbers*!"

It turned out that the number of sinkings was directly proportional to the number of submarines on station and inversely proportional to the number of escort vessels guarding the convoy, *but it was independent of the size of the convoy!*

The answer, of course, was larger convoys. It worked well for a while until the enemy countered by bombing the ports of arrival in the British Isles.

"Another OR problem," an admiral said, "we'd better get some anatomists to work with the RAF."

"None to spare," was the answer, "but we have found entomologists and meteorologists to work with them."

"Entomologists! . . . ?"

"You know, chaps that work with insects, like bees and dragon flies.

Soon, they'll be asking for numbers, so we are warning the RAF to start counting."

The task of mathematics and the other disciplines involved was not so much the solution of specific problems but the nobler and far more important task of bringing man face to face with reality. I believe that the chief virtue of operations research today is that it reduces self-deception and the kind of deception created by eager subordinates who want to please their superiors.

Operations research does not take an exclusively mathematical approach to a problem. One team may be made up of a senior harbor pilot, an assistant harbormaster, two mathematicians, and an engineer. Another could include a highway police captain, a traffic engineer, and two statisticians. A third could be made up of nothing but mathematicians and oil refinery engineers.

The Math of Operations Research

Mathematical tools most often employed in operations research are

statistics and probability
decision theory and game theory
simulation, including the Monte Carlo method
linear algebra and matrix theory
graph theory and the theory of relations

This is not to say that operations research does not use analysis, numerical methods, or any branch of mathematics suited to the problem at hand.

THE MATHEMATICS OF FINANCE

Management decision making utilizes only fractional and limited aspects of the mathematical techniques mentioned above, but the mathematics of finance is an exception. It *is* used by management and can be of great service to value analysis. Remember that the accounting profession contributed the greatest single advance in mathematics since ancient times —the use of Arabic numerals.

There is a story about a naughty but charming little girl who wanted to get into heaven. She ran around the golden walls tossing an apple in the air until it fell inside. Then she asked permission to go in and get her apple. Once she was inside she was so nice to everybody that they let her stay. The mathematics of finance threw the apple of Arabic numerals and algebra into the university campus, where it has grown into a magnificent tree, but it took quite a long time for the owner to come in after it. And even now she is not given true mathematical treatment.

This may well be justified inasmuch as the mathematics of finance has a triple task: to perceive, to inform, and to convince. It exists for a number of conflicting purposes. One is to tell the owner of the business what is happening. Two is to tell the banker and the stockholders that the business is doing very well. Three is to tell the tax collector and the government that the business is not doing well at all.

Not only must the financial manager know what happened yesterday, he also has to know what is happening today and he has to provide information to make it happen better. Such information, in the hands of value task groups, can help the value program to march in step with the rest of the industrial team.

Because of their unique and difficult role in supporting the general manager's "art of the possible," financial interpretations are best left to the experts, but the financial experts have to be brought into the act early enough to guide value analysis in the right direction.

Other mathematical tools, useful in value analysis, are easier to come by. How to acquire them?

A good course of action is to master the particular mathematical techniques required in each problem as it appears. But you have to know what mathematical techniques are applicable; this means you have to know a good bit *about* modern mathematics.

A good way to start is to join the Mathematical Association of America. The dues are moderate, $10 in 1970, and the publications are very good. Then I suggest you go to the library and pick up some of the books in the reading list at the end of this appendix. Start with the elementary ones—they are actually fun to read.

By this time you should receive the *American Mathematical Monthly*. You will enjoy parts of it—there is something in it for everybody—but it is not likely that you will swim deliciously through the contents to surface triumphantly into the sunshine of accomplishment. No, you will probably hit your head on a sunken rock. One of the newer branches of mathematics will have been created, accepted, and permanently established while you were diving.

There is the problem. Which of the new branches of mathematics can you follow? You know you can't follow them all. With how many can you keep up? It may mean giving up golf or being left hopelessly behind. You had been left behind, but you didn't know it. Now it is different.

VALUE-ANALYZE YOUR STUDY PROGRAM

Being a combination of methods for achieving a desired end at minimum cost in resources, value analysis can help you determine what you really want.

The resources in question are spare-time hours and how to use them. This brings into play a complex variable—your wife—and a constant of integration—taking the children to Sunday school.

Even if you can get the variable, within the limits of its range, to cancel out the constant, that is, if you can get her to take the kids herself some of the time, there is really not too much you can do to reduce the expenditure of resources. It is in the upper part of the fraction

$$\frac{\text{desired end}}{\text{cost}}$$

that the gains can be harvested. Your *desired end* was professional improvement. It seemed so important to you that you even considered giving up golf for it. Value analysis asks you these questions, "Is it really that important? Are you twice as good in your job as you are in golf? Would it be three times? You say half! . . . ? You mean you are *better* at golf? Why, as a golf pro you can make a living and you won't need much mathematics—counting up to 150, at worst."

At this point value analysis is being carried away by its rigid objectivity. It is supposed to be objective, except in the area of customer desire. The *desired end* is obviously subjective.

You are the customer. Even if you are better at golf you still want to improve your present or future management skill. Value analysis must accept your final definition of the *desired end*—to learn modern math *and* play golf. It takes another look at your resources to see if this final definition changes matters. No, you still need all the spare time you can get, so it makes its first recommendation: make your wife take the children to Sunday school.

Now that available resources have been determined (Sunday mornings), value analysis tries to convert them into your actual requirements. In your case it is simple. You screen mathematical material for digestibility and economic feasibility, eliminating translations from the Russian in which the verbs are omitted and anything for which the first book costs more than $12. Then you survey the selection and assign to each mathematical technique a coefficient of usefulness on the basis of what it can do for you and a factor of cost on the basis of the units of time required to gain a reasonable understanding of the subject. The higher fractions would, of course, be the better values.

You have to take into account the uncertainty associated with the future. Which of the branches of mathematics will have permanent value and which will fall by the wayside? Will your wife meekly take the children to Sunday school herself or do you have to frame her into being elected a teacher so that she *has* to go? How many friends drop in on you on Sunday mornings?

By this time you will know

(a)   that you can survey n branches of mathematics;

(b)   that the most promising of these are n_1, n_2, and n_3;

(c)   that you will have to frame your wife into being elected a Sunday school teacher;

(d)   that of the friends who drop in on you Sunday mornings one may have to be seriously insulted, but the other two can probably be driven away by warning shots.

At this stage value analysis can help you find the most economical and expeditious path through time and space in order to carry out the study program contemplated in (a) and (b) and to accomplish tasks (c) and (d) in the best sequence and for the least time and effort.

Here is a sampling of other useful mathematical tools at their simplest.

LINEAR ALGEBRA AND MATRIX THEORY

Remember the pleasure with which you solved your first set of simple simultaneous equations? "Neat!" you probably said.

Linear algebra provides the means for putting a lot of information in the form of linear equations and for manipulating the equations to give you more information than you had before—information on previously unknown relations and interactions; for example, to every expected benefit B_1, B_2, . . . , B_n governing the choice of means to accomplish a given end you assign a real, nonnegative number, w_1, w_2, . . . , w_n, which you obtain initially by dividing the number 1 by the number of benefits being considered $(1/B_n)$ and which you then vary as the relative importance of each benefit varies from that of the other benefits.

A positive change in the relative importance of one benefit results in a corresponding negative change in the others. This is imposed by the constraint that *the sum of the measures of relative importance is equal to unity*

$$\sum_{j=1}^{n} w_j = 1.$$

Each of the available choices should meet, in varying degrees of probability or known capability, the requirements of each desired benefit. The measures of the effectiveness with which the choices satisfy these requirements are then transformed from the original units of measure into an interval scale extending from a minimum defined as the least favorable *acceptable* condition to a maximum defined as the best *practical* condition. Such a scale, extending from 70 (minimum) to 90 (maximum) in percentage points has proved to be immediately meaningful to the ordinary business man, accountant, shop foreman, and marketing man who must contribute to the economic aspects of a decision. It is equated by them with familiar school grades, in which 70 is passing and 90 is very good.

Utility and Its Physical Counterpart

The nonlinearity between performance, on the one hand, and the effect of performance, on the other, is taken into account by using a utility function as the means of transformation. You thus use normalization techniques which also adjust for utility and which yield a matrix E, whose elements e_{ij} rate the effectiveness with which the ith choice contributes to the jth benefit.

To arrive at the relative value of the various choices you multiply e_{ij} by w_j because $w_1, w_2, \ldots, w_j$ constitute the measures of relative importance or weighting factors assigned to the desired benefits. This yields the following system of equations:

$$e_{11}w_1 + e_{12}w_2 + \ldots + e_{1n}w_n = v_1$$
$$e_{21}w_1 + e_{22}w_2 + \ldots + e_{2n}w_n = v_2$$
$$\ldots\ldots\ldots\ldots\ldots\ldots\ldots\ldots\ldots\ldots\ldots\ldots\ldots$$
$$e_{m1}w_1 + e_{m2}w_2 + \ldots + e_{mn}w_n = v_m$$

which is what actually appears—in matrix notation—in Figure A-1 and which generates the information v_m from which one selects the highest v_i as the best choice.

The foregoing array may be represented by the matrix equation

$$
\begin{bmatrix}
e_{11}e_{12} \ldots e_{1n} \\
e_{21}e_{22} \ldots e_{2n} \\
\cdot \quad \cdot \qquad \cdot \\
\cdot \quad \cdot \qquad \cdot \\
\cdot \quad \cdot \qquad \cdot \\
e_{m1}e_{m2} \ldots e_{mn}
\end{bmatrix}
\begin{bmatrix}
w_1 \\
w_2 \\
\cdot \\
\cdot \\
\cdot \\
w_n
\end{bmatrix}
=
\begin{bmatrix}
v_1 \\
v_2 \\
\cdot \\
\cdot \\
\cdot \\
v_m
\end{bmatrix},
$$

The weighted ratings add up to relative ·   value

$e_{11}w_1$	$e_{12}\,w_2$	$\cdot\ \cdot\ \cdot$	$e_{1n}w_n$		v_1
$e_{21}w_1$	$e_{22}\,w_2$	$\cdot\ \cdot\ \cdot$	$e_{2n}\,w_n$		v_2
$\cdot\ \cdot\ \cdot$	$\cdot\ \cdot\ \cdot$	$\cdot\ \cdot\ \cdot$	$\cdot\ \cdot\ \cdot$		$\cdot\ \cdot$
$e_{m1}\,w_1$	$e_{m2}\,w_2$	$\cdot\ \cdot\ \cdot$	$e_{mn}w_n$		v_m

which is the product of this vector

w_1	w_2	$\cdot\ \cdot\ \cdot$	w_n

and this matrix

e_{11}	e_{12}	$\cdot\ \cdot\ \cdot$	e_{1n}
e_{21}	e_{22}	$\cdot\ \cdot\ \cdot$	e_{2n}
$\cdot\ \cdot\ \cdot$	$\cdot\ \cdot\ \cdot$	$\cdot\ \cdot\ \cdot$	$\cdot\ \cdot\ \cdot$
e_{m1}	e_{m2}	$\cdot\ \cdot\ \cdot$	e_{mn}

Figure A-1 Matrix theory behind the COMBINEX scoreboard.

where e = effectiveness,
   w = weighting factor or measure of relative importance,
   v = relative value of the ith course of action, or in condensed form,

$$\sum_{j=1}^{n} e_{ij}w_j = v_i \qquad i = 1, 2, \ldots, m.$$

Sensitivity Studies: Perturbing the Weighting Factors

The most delicate, difficult, and important phase of working with an evaluation matrix is the assignment of measures of relative importance. This calls for deep probing into the needs, desires, and values of the customer. How to elicit this information is not within the scope of this appendix, but we can provide a technique for showing the customer, or the requisitioner, the effect that variations in his measures of relative importance, or weighting factors, have on the expected value of the available choices.

This is done by trading back and forth among the weighting factors or by perturbing these factors, one at a time, and studying the effect of this change on the system of equations. The first technique is simple enough, but the second calls for altering the numerical value of *one* weighting factor without affecting the relative proportions among the others. Here is a formula for it.

In the system of linear equations

$$e_{i1}w_1 + e_{i2}w_2 + \ldots + e_{ij}w_j + \ldots + e_{in}w_n = v_i$$

you perturb a given weight, which you identify as w' by changing it to a new weight u and introducing u into the equation in place of the original w'. Then you multiply each of the remaining weights by an expression P, which is $1 - u$ divided by the sum of the remaining weights; thus

$$e_{i1}w_1' + e_{i2}w_2 + \ldots + e_{in}w_n = v_1$$

becomes

$$e_{i1}u_1 + e_{i2}w_2P + \ldots + e_{in}w_nP = v_i'$$

where w' is the weight perturbed,

$$P = 1 - u \Big/ \sum_{j=1}^{n} (w - w')$$

and v' is the new value. Thus one weight can be perturbed while keeping constant the relationship among the others.

Purpose of Matrix Notation

In Chapter 8 we arranged the measures of effectiveness as elements of a matrix and multiplied that matrix by a row vector—all without benefit of matrix notation. Why then do we need matrix notation at all?

The COMBINEX scoreboard of the master plan described in Chapter 8 compared the effect of 185 options on each of four expected benefits—740 multiplications plus 185 additions for a total of 925 operations, not to mention more than 2000 more, had we chosen to perturb one of the weights.

All of these operations can be done by the computer in a matter of minutes. By far the easiest way to program matrix operations into the computer is to use one of the standard mathematical programs of linear algebra. We need matrix notation, therefore, to talk to the computer. It is also the analytical language of linear programming.

Linear programming is a method for assigning, allocating, or combining activities or materials when resources are limited, requirements are specified, and the significant relationships can be expressed as linear equations or linear inequalities. Here is a simple example.

Let's say that the Philadelphia assembly plant of the Mythical Wheeled Toy Company gets its components from company-owned factories located in Chicago (#1), Pittsburgh (#2), and Raleigh, N. C. (#3).

The monthly component requirements of the Philadelphia plant (x_{i1}) can be expressed in the linear equation $x_{11} + x_{21} + x_{31} = 650$ truck-loads (three cells in column one, Figure A-2), read *x one-one, x two-one, x three-one,* where the first subscript indicates the factory of origin and the second, the assembly plant of destination, in this case Assembly Plant

To	ASSEMBLY PLANTS			
From	Philadelphia	Indianapolis	Memphis	Factory capacity in truckloads
Factories	Plant 1	Plant 2	Plant 3	
Factory 1 Chicago	$457	$112	$326	350
Factory 2 Pittsburgh	$183	$212	$455	500
Factory 3 Raleigh N.C.	$240	$380	$449	300
Assembly Plant Orders	650 Truckloads	300 Truckloads	200 Truckloads	1150 truckloads

Figure A-2   The known factors.

#1. The x is the unknown amount that each factory should contribute to make up the 650-truckload requirement.

The dollar figures in the small boxes inside each square show the cost of transportation, per truckload, between the factory on the left of the square and the assembly plant above it. Philadelphia requires 650 truckloads (see bottom of column 1 in Figure A-2). They would prefer to order them from Pittsburgh at $183 freight per truckload, but Pittsburgh can produce only 500 (see right end of row 2 in Figure A-2). Herein lies the problem.

The value of x is governed by the capacity of each factory, the requirements of each assembly plant, and the cost of transportation.

Linear programming provides a framework which organizes all this data for matrix computation in such a way that relationships can be intuitively understood and the problem sometimes solved without resort to linear algebra and matrix theory.

Figure A-2 identifies the known quantities of our allocation problem and relates them to one another. The usual approach to a feasible solution of this problem is illustrated in Figure A-3, where you start at the intersection of row 1, column 1 (the northwest corner), and look for two numbers, one at the end of the row which shows the capacity of the Chicago factory and one at the bottom of the column which shows the requirements of the Philadelphia assembly plant. Transfer the smaller of the two numbers, 350 in this case, to the cell you are considering (1,1). This allocation reduces the requirements at the bottom of the column to 300 truckloads. The next move would be to the right, but you have used up all the capacity in *that* row, so you move downward instead and allocate 300 truckloads of Pittsburgh's production to complete Philadelphia's requirements, leaving 200 to be shipped to other plants. Next, you move to the second column, *second* row, because the first row has been exhausted. You allocate the 200 truckloads left in Pittsburgh to Indianapolis, leaving a requirement of only 100 truckloads which you fill from Raleigh. The 200 tons remaining in Raleigh are then allocated to the Memphis plant, thus completing a feasible solution.

A feasible solution is only a first step. Obviously, Philadelphia would save on freight by ordering from Pittsburgh at $183 a truckload and from Raleigh at $240. No need to order any at $457. Why then the northwest corner rule?

Because in most allocation problems the alternatives are not that obvious, you have to start somewhere; the northwest corner rule at least gets you started. A simple problem such as this one, however, can often be solved by the least-cost rule (Figure A-4) which calls for starting at the least-cost cell and moving from that one to the next feasible, lowest cost sell (Chicago to Indianapolis) until the solution is completed.

To / From Factories	ASSEMBLY PLANTS			Factory capacity in truckloads
	Philadelphia / Plant 1	Indianapolis / Plant 2	Memphis / Plant 3	
Factory 1 Chicago	$457 350	$112	$326	~~350~~
Factory 2 Pittsburgh	$183 300	$212 200	$455	~~500~~ ~~200~~
Factory 3 Raleigh N.C.	$240	$380 100	$449 200	~~300~~ ~~200~~
Assembly Plant Orders	~~650~~ ~~300~~ Truckloads	~~300~~ ~~100~~ Truckloads	~~200~~ Truckloads	1150 truckloads

Figure A-3 The northwest corner rule.

In more complex problems all this scanning, remembering, and comparing takes time and effort. Why should you do all this dog work when computers can scan, remember, and compare? You don't have to own the computer, either. You can lease a time-sharing terminal and use one of the excellent ready-made programs for this and other aspects of linear programming.

Linear programming, as developed beyond the point I have described here, does better than find a good solution. It finds the *best* from among a family of feasible solutions. Following a scientific search pattern which uses each family of solutions as a stepping stone, it provides a guide for determining which way to go, a rule for moving from point to point, and a signal to tell you when you have accomplished optimum allocation.

Not only is linear programming immensely useful, but it has theoretical value in that it combines the synthetic with the analytic tools of

To \ From	ASSEMBLY PLANTS			Factory capacity in truckloads
Factories	Philadelphia Plant 1	Indianapolis Plant 2	Memphis Plant 3	
Factory 1 Chicago	$457	$112 300	$326 50	~~350~~ ~~50~~
Factory 2 Pittsburgh	$183 500	$212	$455	~~500~~
Factory 3 Raleigh N.C.	$240 150	$380	$449 150	~~300~~ ~~150~~
Assembly Plant Orders	~~650~~ ~~150~~ Truckloads	300 Truckloads	200 Truckloads	1150 truckloads

Figure A-4   The least cost rule.

mathematics. The framework of Figures A-2 through A-4 represents the synthetic approach which enabled you to build on the data until the complete structure revealed a first feasible solution. The analytic approach then unravels, takes apart, and examines your first feasible solution in order to arrive at the best possible solution.

In terms of matrix notation, the problem can be stated thus:

Origin	Destination			Capacity
1	$c_{11}x_{11}$	$c_{12}x_{12}$	$c_{13}x_{13}$	$a_1 = 350$
2	$c_{21}x_{21}$	$c_{22}x_{22}$	$c_{23}x_{23}$	$a_2 = 500$
3	$c_{31}x_{31}$	$c_{32}x_{32}$	$c_{33}x_{33}$	$a_3 = 300$
	$b_1 = 650$ $b_2 = 300$ $b_3 = 200$			1150
	Requirements			$\sum_{i=1}^{3} a_i = \sum_{j=1}^{3} b_j$

where c_{ij} = cost of transportation per truckload from each point of origin to each point of destination, as shown in matrix C below,

a_i = capacity of each factory, in truckloads,

b_j = requirements of each assembly plant, in truckloads,

x_{ij} = number of truckloads shipped from each factory to each assembly plant. This is the quantity to be determined.

The c_{ij} above and the matrix C below stand for *cost per truckload*.

$$\mathbf{C} \; = \; \begin{vmatrix} 457 & 112 & 326 \\ 183 & 212 & 455 \\ 240 & 380 & 449 \end{vmatrix}$$

The instructions for solution are find x_{ij} to minimize $\displaystyle\sum_{i=1}^{3}\sum_{j=1}^{3} c_{ij}x_{ij}$

subject to the constraints

$$\sum_{j=1}^{3} x_{ij} = a_i \, (i = 1, 2, 3),$$

$$\sum_{i=1}^{3} x_{ij} = b_j \, (j = 1, 2, 3),$$

where all numbers are nonnegative.

Simple Diagrams from Complex Matrices

If linear algebra and matrix theory are as useful as this section implies, why were they relegated to the appendix? For the same reason that ladies do not wear their girdles on the outside. Factory foremen and executive vice presidents are too busy to digest more mathematics than they need at a given time.

Tell them, "The effectiveness matrix E_{ij} is multiplied by row vector w_j so that all the e_j's in each row can be summed to yield a v_i for each choice . . ." and the foreman will exclaim:

"Now ain't that grand?" in mock admiration.

But you don't notice the interruption. Cold shivers are running down your spine from the cold fury in the vice-president's eyes.

It is much wiser to point to a COMBINEX scoreboard, saying, "Here are the expected benefits. This shows how their importance stacks up. The choices are in this column; the ratings in the middle, and the relative values on the right."

In a value-analysis workshop the COMBINEX scoreboard and the linear-programming framework or tableau go over much better than the analytic approach. The professional value analyst, who must teach others and improve his own methods, however, can profit greatly from linear algebra and matrix theory. For one thing, he will need it when the calculations are extensive enough to be programmed into the computer. Two other mathematical techniques are applicable in value analysis and deserve wider industrial use. They are queuing theory and the Monte Carlo method. Both are briefly described in the section that follows.

QUEUING THEORY AND
THE MONTE CARLO METHOD

The first part of this appendix considered some of the properties of linear algebra and simple matrix theory in order to provide the reader with a scientific yet simple method for comparing alternatives and making allocations. Two other techniques, however, deserve wider industrial use. These are queuing theory and the Monte Carlo method.

To give an idea of the various applications we will start with the selection of a missile-launching system for a small, coast patrol vessel. The ship is destined for close in-shore work either to support or prevent landing operations.

In the "limited war" tasks for which it is being designed the ship will have to have strong air-defense capability. While operating close to shore or in inland waters it will be subject to attack by fighter bombers coming in at tree-top level over the jungle.

We will now study possible launching systems for this hypothetical vessel ordered by the mythical Republic of Freelandia. All our information comes from the unclassified naval experience of Freelandia, whose old coast-defense cruisers and frigates, converted into missile vessels 10 years ago, are too slow and cumbersome. They want a better launching system on a smaller, faster vessel—something smaller than a destroyer-escort but with strong missile capability. See Figure A-5.

System Objective or Function:
Destroy Targets

What targets? Enemy aircraft. What enemy aircraft? Experience shows that a single Freelandia vessel may be subject to sustained attacks by as many as 20 planes. Mean rate of arrival of the planes is one every five seconds.

How well must we do the job? We would like to shoot them all down.

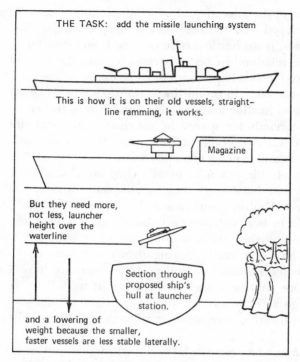

THE TASK: add the missile launching system

This is how it is on their old vessels, straight-line ramming, it works.

Magazine

But they need more, not less, launcher height over the waterline

Section through proposed ship's hull at launcher station.

and a lowering of weight because the smaller, faster vessels are less stable laterally.

Figure A-5 Launching system for a coast patrol vessel.

This has not worked out in practice, but at least we should be able to launch one missile for every attacking plane.

How completely must we do the job to finish the task successfully? We have to launch at least one missile at every attacking plane. It is up to us to decide what it takes to do this in terms of the launching system.

Now, if the enemy planes came in exactly at five-second intervals, we would have no problem, but Freedlandia's enemies are not that predictable. The planes may arrive bunched up and that could be bad. Before we can decide how many missiles we have to launch in any given interval we must know something about the number of planes we will have to contend with during that interval. We have to "queue-up the targets."

SOME ELEMENTS OF QUEUING THEORY

A probabilistic study of telephone-exchange problems led A. K. Erlang (see Brockmeyer, Halstrom, and Jensen) to a study of the consequences of delays and of the allocation of effort to minimize delays. These special studies developed into the theory of queues or waiting lines. A better name, perhaps, is stochastic service systems. If you stop for a moment to think of the relationship between causality and the flow of time, you grasp at once the vast importance of correct sequencing. Queuing theory can be of great value in optimizing certain types of scheduling decision. It is used at oil loading and receiving terminals, in inventory control, in predicting demands for spares, in anticipating the calls for power by randomly activated experiments in spacecraft, in anticipating meteoric bombardment, and in many other activities.

Unfortunately the examples usually cited are ridiculously elementary —customers at a hamburger stand, customers at a newsstand, and cars in a car wash. Unduly ponderous and complex mathematical tools are then brought to bear on these "problems" and the relationship between the formulas and their usefulness becomes more and more tenuous.

People who really need queuing theory and who need it fast use relatively simple formulas on real problems—and we have a real problem here. Our customers are enemy planes. If we fail to "serve" them, they don't go away in disgust, they come right in and *we* get served! The great risk here is congestion or what the enemy would call saturating the target.

We may know how many planes the enemy has and how many he is sending against us, but we don't know, and they may not know, just how or when they will arrive. We may know the average rate of fall of meteors during the Perseid Shower, but we do not know how they will be distributed on arrival unless we have studied the pattern of previous Perseid Showers. Our basic datum is the mean rate of arrival: the total number of events expected or observed, divided by the total time during which they occur.

In ordinary probability we take the total number of events and divide it by the total number of possible occurrences. Not so in queuing theory. We cannot divide all the planes that attack by the total number of planes. There is an infinite number of planes that will not attack us. We cannot divide events by unevents, but we *can* relate events to intervals of time. To do this we need a probability law that not only accepts this relationship but also fits the facts reasonably well.

Such a law is the beautiful, practical, and surprisingly useful Poisson law. All it asks of us is the mean rate of arrival of the expected events and a scale of time intervals. The mean rate of arrival is usually called lambda, mu, U, or m (which I use here). The number of events that happen to fall within a particular time interval is usually called k or x. I use x.

To arrive at such intervals you simply distribute the total time being considered into such regular periods as years, days, hours, or four-second periods, as we do in the following example. The periods do not appear in the table. What the Poisson law reveals is the probability that a certain number of events will occur in any one time period; that is, it tells us if the arrival is likely to follow its mean rate, if you are likely to have no arrivals in that period, or if you are likely to face a congestion during that period.

We can say "The enemy can throw 20 planes at us, each arriving every five seconds. His attack should last $(20)(5) = 100$ seconds. During these 100 seconds the present system can deliver only eight missiles against the 20 attacking planes. Maximum rate of fire is one missile every 12 seconds from two launching arms, each having a 24-second time cycle to load and fire. This inadequate response requires that the vessels travel in company for simple self-protection. No coastline can be watched effectively with the watchmen huddled together in little groups; hence the requirement for small, fast vessels capable of putting up an efficient single-ship defense.

Looking again at Figure A-5, we see that in addition to improving the mechanical performance of the launching system we have to modify its shipboard location. We must increase the height of the launcher while maintaining or lowering the system center of gravity. Now we are dealing with those opposing design requirements described under The Element of Conflict in Chapter 8.

One requirement calls for moving something up and the other calls either for moving everything down or at least for keeping the system center of gravity where it is. Now these requirements exist for entirely different reasons: (a) to improve tactical effectiveness for close in-shore operations and (b) to maintain or improve the ship's stability on the high seas. Before we can make meaningful comparisons we must understand the relative importance of such major design objectives as tactical effectiveness and ship stability. These objectives are listed across the top of the matrix in Figure A-6. The corresponding measures of relative importance, or weighting factors, appear immediately below them.

These weighting factors tell us that increased tactical effectiveness is by far the most important requirement. The 0.5 means that this objective is as important as all the others put together, since they also add up to 0.5, the total sum being unity. Now let us see how we arrived at the ratings under the column headed Increase Tactical Effectiveness, that is, at the numbers 77, 73, 77, and 88 for the choices C_1, C_2, C_3, and C_4. This input of ratings, which appear in the upper left-hand corner of their respective cells, is the output column of the submatrix shown in

Objectives / Weightings / Choices	Increase tactical effectiveness	Reduce power requirements	Improve ship stability	Improve economic effectiveness	Relative values
	0.5	0.1	0.2	0.2	
C_1	77	81	70	76	
	38.5	8.1	14.0	15.2	75.8
C_2	73	70	90	70	
	36.5	7.0	18.0	14.0	75.5
C_3	77	70	82	73	
	38.5	7.0	16.4	14.6	76.5
C_4	88	90	73	82	
	44.0	9.0	14.6	16.4	84.0

Figure A-6   Relating design approaches to their objectives.

Figure A-7. Note that at this stage they appear in the lower half of the cells in the extreme right-hand column.

Putting the Submatrix to Work

The submatrix in Figure A-7 interrelates the elements of tactical effectiveness: cycling speed, launcher height above waterline, readiness, and sector coverage, weighted in proportion to their respective contributions. The proportional measures of this contribution, 5, 3, 1, and 1, should be developed during sea trials and firing exercises aboard an experimental vessel. Far from being subjective measures of preference, these particular weighting factors are based on the physical interaction of the ship, the sea, and the targets.

Tabulated results of these tests assign the greatest importance to cycling speed and to launcher height above the waterline. They also generate a utility function by establishing the relationship between the raw data and the commensurable units we are using to compare ratings; for example, Figure A-8 both normalizes (from the Latin *norma*, a standard measure) and adjusts for utility. The tool for this adjustment is the utility function that models the effect of increased height on launcher efficiency. It reveals that if we were to raise the launcher another deck (for a total of 32 feet) we would run into trouble.

Figure A-9, on the other hand, simply normalizes and does not bother

Objectives / Weightings / Choices	Increase cycling speed 0.5	Increase height above waterline 0.3	Improve readiness 0.1	Increase sector coverage 0.1	Relative values
C_1	76	70	90	90	
	38.0	21.0	9.0	9.0	77.0
C_2	70	70	80	90	
	35.0	21.0	8.0	9.0	73.0
C_3	70	90	80	70	
	35.0	27.0	8.0	7.0	77.0
C_4	90	90	70	90	
	45.0	27.0	7.0	9.0	88.0

Figure A-7   Elements of tactical effectiveness (arrows show direction of ramming).

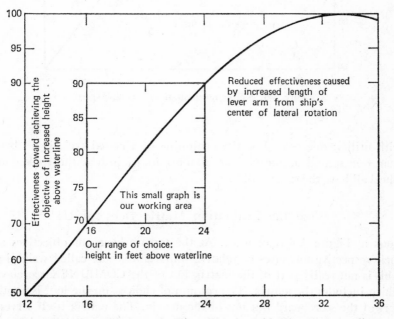

Figure A-8   Normalization and utility adjustment of height above waterline.

307

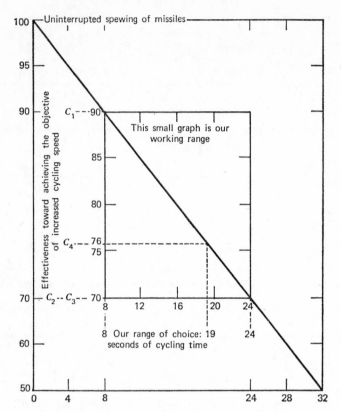

Figure A-9 Simple normalization of cycling speed.

with utility; one second at the beginning of a closed cycle is no better than one second at the end or in the middle, just as the spokes of a wheel all have the same utility.

What the Evaluation Matrix Does for Us

Look at Figure A-6 once more. At the top we have our objectives and their respective measures of relative importance, here called weightings. This is not really part of the matrix but of the COMBINEX scoreboard which includes the matrix. The column of choices, on the left, is another part of the scoreboard but not of the matrix. The matrix itself serves as a tool for relating the means, or choices, on its left, to the ends or objectives above it. The relative value of each choice appears in the output column at the right of the matrix.

The area enclosed by the heavier lines reveals what has been done to improve the original design (C_1). The first attempt (C_2) had been a seat-of-the-pants "improvement" suggested without the benefit of scientific comparison. If one of the objectives was to improve stability, one simply lowered the magazine to the deck below. The launcher remained where it was and stability improved from 14 to 18 in the standard scale . . . BUT something else, more important, suffered. The added reverse-ramming operation plus the increased ramming distance reduced tactical effectiveness from 38.5 to 36.5 and reduced total relative value from 75.8 to 75.5.

Watch the Additivity Assumption!

Let's see what the additivity assumption does to us when the increments of the benefits are not truly additive. Even though our 70-to-90 scale presupposes adequacy or minimum acceptability of all requirements, when we "add" requirements that are partly dependent on each other we get a distorted picture. Substantial differences in relative value are minimized. We still get a fair ranking, but the answers appear deceptively close together, such as the 75.8 and 75.5 in Figure A-6. Unfortunately our objectives or requirements are seldom a true product or a true sum. When they are a true product, we use a different algorithm, but by and large the additivity assumption works well, provided we are aware that apparently small differences in relative value are quite meaningful.

Getting on with the Design

The third choice (C_3) in Figure A-6 also lowers the magazine, with the same disadvantages noted above, but it makes up for it by raising the launcher so that relative value is increased (see Figure A-7). Choice C_4 raises the launcher and lowers the magazine center of gravity slightly to compensate, but its big gains are in cycling time and general simplification. In the other designs the ready missiles and the launcher arms have to meet at a fixed rendezvous and cannot meet at the midpoint between them. In the vertical system, while the launcher is slewing to fire, the magazine can be following it. When the launcher slews to load, the missile can meet its launcher arm at the nearest point. The magazine can index faster because a vertical magazine does not have to handle the weight of an unbalanced load. Now we believe we have a design approach worth working on. It improves relative value from 75.8 to 84.0, a very substantial gain.

This choice could be improved further at the expense of economy. A lighter weight magazine could index even faster, for instance, but this would not raise total relative value. The main thing is that we have improved in the most important area, cycling time, and that this has been accomplished without hurting the other requirements. The 90 under *Increase cycling speed* in Figure A-7 indicates that we have achieved the most favorable condition within our range of choice. Cycling time has been reduced by a full 11 seconds, which gives us an eight-second cycle and a system that can put a missile in the air every four seconds. We now divide the 100 seconds of the attack by the improved four seconds per missile firing to give us 25 periods of four seconds each, or 25 missiles launched. The problem boils down to this:

Twenty attacking planes are distributed over 25 launching periods of four seconds each; $20/25 = 0.8$, which is the mathematical expectation of plane attacks during any one launching period. Obviously whole planes cannot follow a fractional rate of arrival. One, or more, or none at all, will arrive during each launching period. *That* is the variable of interest; we call it x. Say in one period x will equal 0 arrivals, in another it could equal 1, and in certain periods it might equal 2 and even 3. What are the chances of 0, of 1, and of 3 arrivals?

If the planes follow the Poisson law under parameter m (Parzen, 1960), the answer is given by the distribution function $e^{-m}m^x/x!$ where $x =$ the number of events in any one period, $m =$ the mean rate of arrival.

The probability of $x = 0$ arrivals, $x = 1$ arrival, $x = 2$ arrivals, . . . , $x = n$ arrivals can be found for limited values of m and x on page 259 of Burington and May (1959 edition) and on page 444 of Parzen (1960). Note that Parzen uses λ for our m. Entering the tables with the number of estimated arrivals x and the mean rate of arrival 0.8 we get the results tabulated in Figure A-10.

We note at once that there are 3.60 periods, each subject to a two-plane attack.

A GLIMPSE OF THE MONTE CARLO METHOD

Gamblers who have no time to take down all the numbers that come up in roulette over two or three months and who want to develop a winning system can buy a table of random numbers, work from that, and give up without losing their money. The Monte Carlo method is useful in a surprising number of ways, such as the integration of irregular areas when the boundary can be broken up into known functions.

A frequent and useful application of Monte Carlo is simulation of the

SYSTEM OBJECTIVE: Defend against sustained air attack by 20 planes.

DATA FROM COMBAT EXPERIENCE: Sustained attacks by 20 planes last about 100 seconds. Mean rate of arrival of attacking planes is one every five seconds.

SYSTEM CAPABILITY: Eight-second, alternate cycling time from dual launcher can put a missile in the air every four seconds.

COMPUTATION: One hundred seconds of attack divided by four seconds to launch a missile

$\qquad$ = 25, the total number of four-second "service" periods

$\qquad$ = T (total periods).

$\qquad$ Number of estimated "arrivals" = 20 = N.

$\qquad$ N/T or 20/25 = 0.8 = estimated parameter m, where

$\qquad$ $e^{-m}(m^x/x!)$ is the probability that x arrivals will take

$\qquad$ place during any four-second period.

$e^{-0.8}\left(\frac{0.8^x}{x!}\right)$	x	Multiplied by 25	
Probability that...	So many planes will attack in any five–second period	Applying these probabilities to the total number of periods, we get the number...	
		of planes	in so many periods
0.4493	$x = 0$	0	11.23
0.3595	1	1	9.00
0.1438	2	2	3.60
0.0383	3	3	1.00
0.0077	4	4	0.20
0.0012	5	5	0.03
0.0002	6	6	0.00

Figure A-10   Queuing the targets.

type, "What would happen if . . . ?" which identifies beforehand the role likely to be played by chance in a hazardous situation; for example, we are worried about those 3.6 periods in which our little ship will have to fight off two planes when it can send up only one missile.

We would like to know how closely the deadly two-plane attacks are likely to bunch up, how closely the almost fatal three-plane attack will come to one of them, and how the more manageable attacks, and the slack time, will be distributed throughout the 25-second raid.

Somebody says, "Let's run a Monte Carlo on it."

The Nature of Random Numbers

Utilizing the element of chance for our own ends is the essence of the Monte Carlo method. To find out what can happen when chance takes over we seek out the unpredictable and harness it. An early part of this harness was the martingale or checkrein used to describe a betting system which can be defeated only by limits on the size of bets or the time of play. Other early tools for harnessing chance were the ritual objects of the Goddess of Fortune: dice, cards, and the roulette wheel, hence the name Monte Carlo method first used by Metropolis and Ulam.

Dice are probably the oldest random number generators. The RAND Corporation's *A Million Random Digits* is one of the most modern sources. In this example, however, we shall continue to use the excellent short table from the CRC Standard Mathematical Tables because these tables are so readily available.

Now, to simulate the range of probability from 0 to 1 we need 10,000 numbers in four digits. We have this in the series 0000 through 9999 because the entry 0000 plus the entries 0001 through 9999 together add up to 10,000. The stand-in for 10,000 is the entry 0000. Using r to represent any random entry in this series, and x to represent the number of planes that can attack in any one period, we develop the simulation model shown in Figure A-11.

The chances that no (0) planes will attack during a given four-second period are 0.4493; so we assign all numbers greater than 0 and equal to or less than 4493 to represent $x = 0$ (no planes attacking).

The chances of single-plane arrivals are 0.3595. We add the number 3595 to 4493 to get 8088, as shown in the model. Then we assign all random numbers greater than 4493, but not greater than 8088, to simulate the chances of a single-plane attack ($x = 1$).

The same procedure is followed until we reach the 12 numbers greater than 9986, but not greater than 9998, which simulate the very remote probability (0.0012) that four planes will attack during a given four-

CHANCES ARE SIMULATED BY A
NUMBER (r) DRAWN AT RANDOM

THE CHANCES		WHEN r IS GREATER		BUT SMALLER OR EQUAL	
THAT	ARE	THAN		TO	WRITE
$x = 0$	0.4493	0001	$\leq r \leq$	4493 +3595	0
$x = 1$	0.3595	4493	$< r \leq$	8088 +1438	1
$x = 2$	0.1438	8088	$< r \leq$	9526 + 383	2
$x = 3$	0.0383	9526	$< r \leq$	9909 + 77	3
$x = 4$	0.0077	9909	$< r \leq$	9986 + 12	4
$x = 5$	0.0012	9986	$< r \leq$	9998 + 2	5
$x = 6$	0.0002	9998	$< r \leq$	0000 (10,000)	6
$\sum_{x=0}^{x=6}$	1.0000	where 0000 represents the 10,000th random number			

Figure A-11 A model for Monte Carlo simulation.

second period. Now only two numbers, 9999 and 0000 (representing the number 10,000 which cannot be shown in a four-digit table), are all that is left to simulate the 0.0002 (two in 10,000) chance of a five-plane attack ($x = 5$).

Simulating an Air Attack

In 20 out of the 25 "service periods" either one plane or no planes at all arrive for "service." A service period, in queuing theory terms, is simply the time it takes to serve a customer in the queue. Remember it takes our launching system four seconds to launch a missile from dual launching arms, each having an eight-second cycling time. So it takes us four seconds to "serve" a customer, that is, to launch a missile at him. We are preparing the defense against 20 planes whose sustained attacks last an average of 100 seconds. It takes us, therefore, 25 four-second periods to serve 20 customers in 100 seconds.

The numbers suggest 20 out of 25 instances of adequate service. It would be nice to know, however, what happens in the five less comfortable periods. So we simulate a succession of 25 four-second periods in each of which we can launch only one missile.

Now for the simulation. We select a column at random from a table of random numbers, say column 7 on page 238 of the CRC Handbook. To keep the example simple we run straight down the column and copy the first four digits of each number (see Figure A-11); for example, 6917 is greater than 4493 but smaller than 8088, so we write 1 (one plane attacks during the first four-second period). The next number, 2798, is less than 4493; we write 0 (no planes attack during the second period). The 25 periods of the attack follow this pattern: 1 0 0 0 1 0 1 2 1 0 1 0 0 1 1 1 1 0 2 1 2 1 2 1 1.

In a real simulation, of course, we would do many more iterations. This is the sort of thing in which computer time vastly increases the reliability of the Monte Carlo method. It is remarkable, though, that in this minimal sample the random numbers simulated the total number of enemy planes with 95% accuracy. None of the two-plane attacks come together and there is not one single three-plane attack. Such a brief simulation could lead to disaster. Let the reader continue the simulation for a number of 25-period attacks. He will note that the third attack (line 5, column 7, page 239) begins with a deadly cluster:

8303 simulates 2 planes
9766 simulates 3 planes!
8882 simulates 2 planes

and during this seven-plane attack the ship could launch only three missiles!

Now we understand the function. Destroy seven targets in 12 seconds. This is the difference between seat-of-the-pants value analysis and value analysis with numbers. The missile launcher we are analyzing can launch only three missiles in 12 seconds. This is not so bad as it sounds because the function of destroying seven targets in 12 seconds applies to the whole air-defense system, including guns. This function must be factored to take into account the fact that some of the attacking planes will miss the ship altogether. Still, we must launch at least five missiles in 12 seconds to provide the function of the launching system itself. We can really pour in the money and reduce the cycling time—sophisticated bearings in the massive magazine, light metals and aircraft-type structures to reduce inertia; more power for indexing the magazine, for transferring missile to loading rail, for operating spanner rails and blast doors, for ramming and retracting the ram-chain and pawl, and for slewing to fire and slewing to reload.

Or we can stop to think.

If we maintain the present level of performance but minimize cost by simplification, innovation, and updating, we can have two launchers, one forward and one aft. These two simple and rugged systems will deliver six missiles in 12 seconds for the same money as a single more sophisticated system tuned to the quick. The two simple systems will do the job more reliably and can be delivered sooner.

The Monte Carlo Method and Random Sampling

It is advantageous to distinguish between random sampling, systematic sampling, and stratified sampling on the one hand and the Monte Carlo method on the other. Sampling theory has broad applications (see Snedecor) and in its more advanced forms requires sophisticated calculations. The Monte Carlo method is more limited in application, but sometimes it is the only way to get the answer (Shreider), and it constitutes a fascinating, exacting, and often indispensable craft (Hammersley and Handscomb). Other good sources on Monte Carlo are Householder, Forsythe, and Germond; also Meyer. Alwin Walther's paper in Meyer, page 278, is a delightful introduction: ". . . two girl students are called up as goddesses of fortune. . . ."

Other Good Mathematical Books

Kemeny, Schleifer, Snell, and Thompson is a good compendium of modern mathematics for managers; so is Levin and Kirkpatrick. Deeper and more general works on the mathematics of resource use are A. D. Hall, Hare, Hillier and Lieberman, and Churchman, Ackoff, and Arnoff.

Statistics. Hoel's is the kind of a book you can understand as fast as you can read. Snedecor is also very readable and extraordinarily complete.

Probability. My three favorite books on probability are Fry (1965), Parzen (1960), and Feller (1950). Feller's Volume II (1966) is very advanced, as is Parzen (1962), but both are sometimes necessary.

Linear Algebra. A delightful and truly introductory book on linear algebra is that by Stewart. This book is an excellent example of the art of teaching as distinguished from the craft of presenting knowledge. A somewhat deeper, more rigorous, and less introductory introduction is Marcus and Minc. It is a good second volume to follow Stewart's.

Matrix Theory. MacDuffee (1943!) is an excellent little book, very good as a starter. Somewhat more general and quite readable are Thrall and Tornheim; also Hohn.

Linear Programming.   For crystal-clear exposition, sound theory, and useful charts and diagrams Loomba is hard to beat. If I could buy only one linear programming book this would be the one. Garvin, Vasonyi, and Ferguson and Sargeant are all reasonably easy to use and theoretically sound.

Queuing Theory.   An excellent book on this tricky subject is Riordan. Khintchine, Saaty, and Cox and Smith are other good, specialized books. Both of Parzen's books, cited under *Probability,* above, are of great help; the 1960 book, pages 251 to 258 (*queuing theory* is not mentioned in the index!), and the 1962 book which includes one of the most complete studies of the Poisson process offered to date. The operations research books listed below all offer worked-out problems on queuing theory, although it may be listed in the index under *waiting lines.* The examples in Churchman, Ackoff, and Arnoff are particularly readable.

Statistical Decision Theory.   Grayson, Ackoff, and Churchman (1961) are practical books that can be used immediately without going back to school. The reader gets the benefit of the vast experience and scholarly accomplishments of the authors, but these trappings do not show. Somewhat more theoretical are Chernoff and Moses, and Schlaifer. The most profound, thorough, and scholarly work to date on the mathematical aspects of value and decision theory is Fishburn. The giants, on whose shoulders these writers stand, are Von Neumann and Morgenstern, and Savage

Game Theory.   Williams is a fun book which turns out to be full of solid, applicable material. It is a good starter. The great classic, of course, is Von Neumann and Morgenstern mentioned above. McKinsey and Luce and Raiffa are excellent working books.

Operations Research.   My old favorites are Churchman, Ackoff, and Arnoff, and Sasiene, Yaspan, and Friedman. An invaluable book for business managers is Hillier and Lieberman.

Appendix B

MEASUREMENT

Measurement is one of the notions which modern science has taken from common sense.

Norman Robert Campbell

The common sense origin of measurement is evident in the industrial measurement of manual work but not of knowledge work. Many of the measures applied to industrial departments have little to do with common sense, let alone science. I refer to measures that are *un*incentives and to measures that reward a department for looking good on the books at the expense of other departments. How to avoid this pitfall in measuring a value program is discussed after we define measurement, itself.

Before attempting to define measurement we should consider the functions of definition. Beardsley, in his *Practical Logic*, assigns two functions to definition: *supply meaning* and *restrict meaning*. At a lower level of abstraction Schiller stated that a definition must serve a given purpose by making it convenient or possible for the term defined to state the essence of what is important at the time. We can describe Schiller's function of definition as *state essence*.

DEFINITIONS OF MEASUREMENT

Campbell (1921): "The assignment of numbers to represent properties," a definition that states the essence of measurement.

Bross (1953): "Classification by numerical values," another definition that states essence.

Stevens (1959): "The assignment of numerals to aspects of objects or events according to rule," a definition that states the essence and supplies the meaning of measurement.

Ackoff (1962): "A way of obtaining symbols to represent the properties of objects, events, or states, which symbols have the same relevant relationship to each other as do the things which are represented," a definition that states essence, supplies meaning, *and* restricts meaning. Ackoff bases it on this reasoning

> We can define measurement as a process whose output can be used in a particular way. This makes the essential property of measurement the *functional* property of its product, rather than the way it is produced. . . . Therefore, in general terms, *measurement can be defined in terms of its function.* (The italics are mine.)

To gain the full benefit of Ackoff's reasoning we can include his comments in the definition itself (Ackoff, 1962)

> *Measurement*: a process whose output can be used for obtaining symbols to represent the properties of objects, events, or states in such a way that the symbols have the same relevant relationship to each other as do the things they represent. The essential property of measurement is the functional property of its output.

THE FUNCTIONS OF MEASUREMENT

"It is easier to send you the measurements than it is to send you the girls," wrote my brother Eduardo from Bogotá. "Anyway, you will be in New Orleans, where girls are abundant; so all you need are the measurements. Just buy a nice dress for each girl and send it to her at the Posada de la Lora near the Quindío pass on the Western Cordillera."

My brother Eduardo was invoking the *Principle of Economy.* An experienced seaman, he learned early in life that it is more economical to measure a truck to see if it will fit through the hatchway of a ship than it is to hoist it over the hatchway, shout, "Lower away!" and let the truck find out for itself.

It is quicker and easier to compare the measurements of objects than it is to bring the objects together for comparison. It is also quicker and easier to compare the measurements taken of events, states, and processes than it is to reproduce the conditions and results. Cooking from a recipe is an example. It may be less satisfying, but it is certainly more practical than recreating the right combination of ingredients, processes, time, and temperature that gave rise to the original dish; more important, the dish can be reproduced by someone less gifted than the original creator. Here are some of the main functions of measurement.

CONSERVE RESOURCES. This function implements the *Principle of Economy*. Measurement helps to save time, conserve space, eliminate excess weight, conserve energy, and reduce human effort.

IMPROVE PRECISION. Precision means degree of exactness. It is possible to hit the exact center of a disk with an arrow, even if the center is not marked, and still achieve 100% precision, but, as Plato put it, it would be a matter of "guesswork and lucky shots." Measurement makes precision reproducible. Costly measurements can increase precision and less costly ones often reduce it. All these are matters of fact. Improving precision, on the other hand, is a matter of value. It means achieving better precision for a given purpose.

IMPROVE ACCURACY. Accuracy means freedom from error or degree of approximation to the truth. Itself a measurement, accuracy measures how closely other measurements come to the truth. Great accuracy is costly. Poor accuracy can be economical but also disastrous. The right degree of accuracy is a matter of value. It depends on "the uses to which the measurement is put" (Churchman 1959).

CONTRIBUTE REPEATABILITY. Thanks to measurement the same experiment can be done elsewhere by other persons to verify the principles discovered. When probabilities are a major input, we can invite the Goddess of Chance to play with the other measurements and we have simulation; that is, we can learn the results of certain *other* events by simply changing the odds. Repeatability is, of course, the essence of mass production.

QUANTIFY PERFORMANCE. Payload, speed, ceiling, and range of an airplane, horsepower of an engine, box-office receipts from a play, and copies sold of a book show how measurement describes performance in numbers.

MONITOR PROGRESS. This is the function of measurement that athletes call "pacing yourself," teachers call "grading the students," and bosses call "measuring performance." It is discussed at the end of this appendix.

There are many other functions of measurement, but the ones listed here are the most important in value analysis. Here they are, in summary:

CONSERVE RESOURCES
IMPROVE PRECISION
IMPROVE ACCURACY
CONTRIBUTE REPEATABILITY
QUANTIFY PERFORMANCE
MONITOR PROGRESS

SCALES OF MEASUREMENT

A nominal scale assigns numbers or other symbols, using them as *names*, to establish both identity and difference among units. There are two rules: all symbols are different and only one symbol to a customer. The numbers on football players, car license plates, police badges, and credit cards are examples. Campbell (1921) has this to say about the nominal scale

> Numerals are also used to represent objects, such as soldiers or telephones, which have no natural order. They are used here because they provide an *inexhaustible* series of names, in virtue of the ingenious device by which new numerals can always be invented when the old ones have been used up.

The nominal scale preserves identity and difference but not order, interval, or ratio.

Note. The term *units* is used in this section for units, events, or states.

The ordinal scale tells us which unit comes before the other, as house numbers do. House numbers describe the sequence in which houses are *ordered* along the street, but they do not tell us how far apart the houses are nor how wide each house is with respect to the others. Mohs' hardness scale is an ordinal scale. The ordinal scale preserves order but not identity, difference, interval, or ratio.

The interval scale uses uniform intervals to indicate the location of units along a line. Streets or avenues, when numbered or lettered in regular sequence, constitute such a scale. It not only tells us that 64th Street comes after 60th Street and before 68th Street but it also tells us how far. It measures the interval. Passing grades in school are usually scaled along an interval scale ranging from 70 to 100. The calendar is an interval scale. The interval scale preserves order and interval but not identity, difference, or ratio.

The ratio scale is characterized by regular intervals, sequentially numbered from a known zero point. It presupposes that the observer "knows" where true zero is, as in the Kelvin and Rankine absolute temperature scales. Twelve inches, 60 minutes, 360 degrees, and 100 cents are ratio scales. If we line up the end points, we find that 6 inches, 30 minutes, 180 degrees, and 50 cents coincide because each is one-half of the whole and the numbers themselves bear the same ratio: 6:12 as 30:60 as 180:360 as 50:100. Three inches, 15 minutes, 90 degrees, and 25 cents each represent one-quarter of the whole. Three-quarters is 9 inches, 45 minutes, 270 degrees, 75 cents, and so on. Within each scale the numbers on the scale bear the same relation to one another as the units they measure. The ratio scale preserves order, interval, and ratio but not identity or difference. It tells us that a police officer, weighing 300 pounds, is heavier than one weighing 200 pounds (*order*); it tells us how much heavier he is (100 pounds, the *interval* between 200 and 300) and that he weighs half again as much as 200 pounds (*ratio:* 200 is to 300 as 1 is to $1\frac{1}{2}$), but it does not tell us *who* the 300-pound officer is nor how to tell him from another 300-pound officer. His badge number *does;* it preserves identity and difference but that is all.

The bounded interval scale deserves a note of its own. It is an interval scale with lower and upper bounds that limit the range chosen for measurement. These bounds exclude irrelevant information and improve signal-to-noise ratio. In this sense the bounded interval scale can serve as a screening device, excluding useless quantities, such as *too little* and *too much*. It is particularly useful in local comparisons, within the range chosen for measurement, in which the absolute minimum and maximum of the quantities involved is of no interest.

SOURCES ON SCALES OF MEASUREMENT

Campbell (1957) is the classic on modern measurement. His system depends on an initial distinction between fundamental and derived measurement which requires an analysis of measurement procedures. Following Ackoff's approach to measurement, we are more interested in the functional output of measurement. S. S. Stevens' classification of scales, as modified by Coombs and by Ellis, is based on what the scales can do for us. Stevens, Coombs, and Ellis, therefore, are the main sources for my classification of scales.

MEASURING A VALUE PROGRAM

An industrial improvement program may find an ideal territory—rich and unoccupied—and still not get off the ground. Resources must be committed to exploiting the territory, and such resources will not be forthcoming if they can be invested more profitably somewhere else.

We can insist that value analysis is bound to produce better results, and management may well act on our opinion—just that once—to see the numbers we come up with, but unless we can support a continuing program with well-audited figures we are offering only opinion, and management seldom acts on opinion when facts are available.

Operating Figures

In addition to their formal accounting and auditing procedures, the controller's people have to be skilled in quick horse-sense appraisals of worth and cost. They can be called on to provide operating figures that identify the actual dollars, man-hours, and machine hours saved by an improvement activity when such dollars, man-hours, and machine hours are put to a profitable use.

Operating figures exist for other purposes, of course. They are developed by accounting and estimating personnel for real-time control as distinguished from after-the-fact reporting.

Surprisingly enough the present and the future are easier to put down in numbers than the past. The past is certain and rigid and has to add up or we have made a mistake somewhere.

We really never know for sure what is happening right now, and much less what *will* happen! We come up with approximate figures and nobody expects them to add up.

Figures based on the past are accurate and useful; figures to interpret the present and forecast the future are less accurate but more useful. Hence the advance in the accounting profession: from accounting for the past to providing guidance for the present and the future.

So, with the help of Accounting, Manufacturing Standards, Cost Estimating, and the line departments involved, we develop worth figures for what value analysis actually accomplishes over a given period. To show what we have to watch out for I will describe one pitfall.

Warning: Paper Savings!

The operating figures must exclude such paper savings as dollars "saved" in one area which increase cost in another, machine hours "saved" on a machine that is idle half the time, square feet of storage space "saved" in an oversize building, and labor saved in operations in which the operator must remain at his post and the product does not move any faster through the assembly line.

Generating savings that do not save steals time from real savings and destroys confidence in any improvement program.

Hard-Dollar Savings

Real savings must come out of money that would have to be spent under conditions of good design, good buying, and good manufacturing. Such money must exist; it must form part of an honest and conscientious estimate, work order, quotation, or actual cost; the money must *be* somewhere and the savings action must generate a transfer of funds from where the money is to where it can be used more profitably.

Once we have a method for auditing and transfering savings and operating figures for calculating our own cost we can value-analyze the program.

Worth of the Value Program

Here is the typical year-end report of a manufacturing plant:

Savings from Value Analysis

48 single product tasks	$2,803,645
5 plantwide workshops	1,118,945
Net savings	$3,922,590

after deducting such costs of implementation as task-group time, engineering time, drafting time, materials, tools, testing, idle time for changeover, and engineering follow-up.

Cost of the Value Program

Value Analysis Operating Cost

	Single Product Tasks	Plantwide Workshops
Salaries	$47,277	$ 30,623
Phone, telegraph, and cable	3,015	2,110
Travel	5,493	3,486
Research and development	11,615	7,381
Facilities	8,313	5,322
Supplies	687	478
Staff support	2,338	4,852
	$78,738	$ 54,252
Total cost of value analysis		$132,990

Value of the Program
Computation of Value

$$\frac{\text{Worth} \quad \$3,922,590}{\text{Cost} \quad \$ \ 132,990} = 29.5 \text{ index of value}$$

Measure of Gain or Loss

Worth	$3,922,590
Less cost	$ 132,990

= $3,789,600 gained

Warning: The Numbers Game

Twenty-nine to one! What a return! . . . Caution . . . caution . . . caution! The $3,789,600 is less than 4% of the plant's $30,000,000 sales, but 4% *added to profit!* . . . Caution . . . caution . . . caution! The 4% extra profit is the work of everybody in the plant. Value analysis simply helped to guide it through the difficulties created by rapid growth, increasing complexity, frozen information, and poor communications. Pulling the right log out of a log jam may get the whole mass of timber moving again, but the lumberjack who moved that log doesn't claim credit for cutting the timber. Similarly, taking sand out of the gears of an engine may yield a return out of all proportion to the cost of cleaning the gearbox, but no amount of cleaning generates horsepower. The gains belong to the risk taker who stopped the engine long enough to have it cleaned.

Monitoring performance of value analysis therefore is not a matter of comparing the dollar gains of the value projects with the dollar cost of the value effort but of monitoring the total effect of the value effort on *all* plant operations. If the general feeling in the plant is that value analysis is good for everybody and helps make a profit, then value analysis is on the right track, even though its dollar gains may be modest.

If, on the other hand, the general attitude toward value analysis is one of suspicion and hostility, then value analysis is doing something wrong. It may be saying all the right words but it may be taking the wrong actions—embarrassing people, showing them up, disrupting schedules, or simply mouthing pious platitudes, such as revealing to captive audiences the great truth—presumably unknown until that moment—that the company is out to make a buck.

The best performance measure of value analysis is the number, frequency, and insistence of requests for its services by departments other than the one that sponsors it.

Appendix C

PROFESSIONAL FOUNDATIONS

Profession: A calling or vocation requiring specialized knowledge.

Saying that a substantial body of knowledge underlies the value disciplines sometimes raises the hackles of those down-to-earth value specialists who have little patience with academic language. They may be right too, so I will describe a value improvement effort in good factory-floor language, avoiding value terminology. We can then see if there is any real knowledge applied in the value effort. Here is the job pl— . . . oops! Here is the way we do it.

1. GETTING READY
   Picking out what to work on
   Putting the right skills in each group
   Turning the groups into teams
2. GETTING THE FACTS
   What does it cost us to make the product?
   What does it cost the customer?
   What is it worth to the customer?
3. WHAT DOES THE PRODUCT OR SERVICE DO?
   Of all it does, what's most important?
4. WHAT ARE BETTER WAYS TO DO IT?
   Inventing on schedule
5. WHAT WILL EACH IMPROVEMENT BUY US?

What will it cost us?
What is the risk?
Which is the best idea?
6. TELLING OUR BOSSES
What we think ought to be done
Why it should be done
7. GETTING IT DONE
Who will do it?
Where?
How will it be charged?
When will it be done?
How long will it take?
Who checks up on it?

There it is; as simple as I could put it. To make sure I was using down-to-earth language I read it to my five-year-old granddaughter.

She stopped me at the words: skills, group, product, service, and customer, which I duly explained.

"Now do you understand the whole thing?" I asked her.

"Yes, Grandpa."

"Do you think you could walk into a factory and do all this?"

"What's a factory?"

So you see, understanding what is to be done is not enough; some additional knowledge is implicit in the job plan. No matter how simply we try to put it, it is not kindergarten stuff.

Our rugged, down-to-earth value specialist exclaims, "Shucks, I know what a factory is. I know who to talk to to get things done. I can set up a value workshop without the help of books, too!"

True, and if he is at all successful he knows a great deal more. You can see it by looking at him. He is covered over with the scars of experience, of learning the hard way. He has reinvented numberless wheels until he found out, in the school of hard knocks, that the round ones are best. But we all know dozens of value specialists who flunked out of the school of hard knocks, who never made it.

We need a true discipline and a better learning procedure. Now the word *discipline* comes from *discipulus*, a pupil or learner. The abstract noun, such as in "he maintains good discipline," means obedience, but the concrete noun, as in "economics is a business discipline," simply means *a subject that is taught: a field of study*.

The great diversity of knowledge applied by the successful value specialist cannot be taught in a two-week seminar; but it can be taught, as the professions are taught, in the university.

I shall try to identify the branches of knowledge applied by our rugged, down-to-earth value specialist; but I shall do so in academic terms—the factory-floor language of the university.

In the list below the (U) following a subject indicates that it is already taught as a regular course in most universities, the (+), that it could be added to the current curriculum for the benefit of most students, and the (V) designates specific value improvement subjects. The number preceding each paragraph relates it to the areas of effort listed above. The title is the corresponding step in the job plan I use most often.

1. PREPARATION. Planning Theory (+), Economics of the Consumer and the Firm (U), Interaction of Industrial Disciplines (+), Introductory Social Psychology (U), Group Dynamics (U).

2. INFORMATION PHASE. General Information Theory (V+)* Cost Accounting (U), Theory of Value (V+), Business Statistics (U), Industrial Purchasing (U).

3. ANALYTIC PHASE. Product Dynamics (V)†, Other possible titles for this subject are Function and Structure, or Performance Evaluation.

4. CREATIVE PHASE. Thinking and Problem Solving (U), Product Innovation and Marketing (U), Theory of Design (U), Mechanics of Materials (U), Human Factors Engineering (U).

5. EVALUATION. Managerial Economics (U), Evaluation Methods (V+), Estimates and Costs (U), Utility and Consumer Choice (V+), Marketing Methods (U), Economic Decision Making (U).

6. PRESENTATION. Professional Expression (U), Techniques of Presentation (V+).

7. IMPLEMENTATION. Psychology of Motivation (U), Budgets and Budgeting Control (U), Engineering Administration (U), Manufacturing Processes (U), Manufacturing Equipment and Materials (U), Work Design (U).

---

* Nearly all information theory is really communications theory. The value disciplines *have* developed methods for gathering and appraising information.

† This is the vital aspect of value improvement that concerns itself with the function rather than the structure of a product. "Functional analysis" is a subject listed in university catalogs under advanced mathematics, usually in the graduate school. The University of Wisconsin requires 17 semester credits in mathematics *beyond the calculus* before you can take functional analysis. Even *function analysis,* which is readily understood in an industrial environment, causes confusion in a listing of college subjects, in which the words *function* and *analysis,* used together, at once suggest mathematics.

THE SUCCESSFUL VALUE SPECIALIST

The full-time value specialist, whose career reflects progressively increasing responsibility, remuneration, and results, must of necessity apply the branches of knowledge listed above. An examination of job descriptions and recruiting specifications reveals that this body of knowledge is a condition for employment of value analysts and value engineers.

The requirements are boiled down into such paragraphs as the following:

He must interact well with peers and superiors, effectively influence others, and be able to organize and direct interdiscipline task groups. He must also find ways to motivate engineers to change their designs, buyers to change their purchase orders, and department managers to accept and do unscheduled work. This man must understand the creative process and be able to stimulate inventiveness in others (Psychology).

He must show sound, business horse-sense, think in terms of the company and its products, and have a thorough understanding of costs and profits (Business Administration, Economics, Accounting).

To work across the board in an industrial plant the value specialist must understand the technical language of key departments, be familiar with their particular procedures, and with the pattern of their interaction. He should be able to present recommendations in both business and technical terms, providing management with brief bottom-line answers (Industrial Operations).

COMBINING THE INGREDIENTS

Now we must put these personnel requirements into college recruiting terminology. Here is what we get:

Table C-1

Subject	Credits
Introductory Social Psychology	3
Psychology of Motivation	3
Group Dynamics	3
Thinking and Problem Solving	3
Product Innovation and Marketing	3
Marketing Cases and Problems	3
Business Statistics	3
Economics of the Consumer and the Firm	3
Managerial Economics	3
Economic Decision Making	2

Table C-1 (*continued*)

Subject	Credits
Cost Accounting	2
Estimates and Costs	2
Budgets and Budgeting Control	2
Theory of Design	2
Human Factors Engineering	3
Mechanics of Materials	2
Manufacturing Equipment and Materials	2
Manufacturing Processes	3
Work Design	3
Industrial Purchasing	2
Professional Communication	3
	55

These credits represent the college subjects listed initially under the seven headings of a minimal value improvement effort. The courses are all taught in college today under the exact designation and with the credit hours shown in the list.

Such is the special combination that the value disciplines abstract from the general body of knowledge. In the same way, however, that Mechanical Engineering must add such subjects as Mechanical Design, Engineering Graphics, and many others to the Physics, Chemistry, Mathematics, and Economics that it abstracts from the general body of knowledge, so must the value disciplines add their own special subjects.

Table C-2 Value Improvement Subjects

Subject	Credits
Planning Theory	2
Interaction of Industrial Disciplines	2
Theory of Value and Valuation	3
Principles of Value Analysis	3
The Value Analysis Method	3
General Information Theory	3
Product Dynamics	2
Utility and Consumer Choice	2
Theory of Creativity and Practice of Innovation	2
Measure Theory	2
Mathematics of Resource Use	3
Contracts and Specifications	2
Economic Selection	2
	31

CURRICULUM

The value specialist, therefore, applies in his daily work professional knowledge that is now offered in regular university courses. He also applies the specialized knowledge peculiar to his own field. In terms of curriculum planning the subject matter adds up to this:

Value improvement courses already being taught	55 credits
Specialized value improvement courses	31 credits
Total credits in value	86

This number is right in the ball park with 76 to 86 credit hours of engineering subjects taught in a four-year course leading to a bachelor's degree.

In addition to the credit hours in the major field, an engineering course provides credit hours for such universally required subjects as Freshman English, Mathematics, Physics, and Chemistry, and a number of electives.

	EE	ME	IE
Engineering subjects	76	86	84
English, Mathematics, and Basic Science	39	26	27
Electives	25	24	26
	140	136	137

The course material identified so far corresponds exactly to the Mechanical Engineering distribution: 86 credit hours in value improvement, 26 basic freshman credit hours, and 24 electives, adding up to 136 credit hours for a four-year course. Just what are electives?

ELECTIVES

The subjects characteristic of a particular discipline are usually supplemented by closely related subjects that cover special applications. In education these supplementary subjects form part of a group of "electives" which a student can use to meet his particular needs.

When we have to learn a discipline the hard way, we do not *choose* the electives; they happen to us. The value specialist in a small plastics firm is told to work with Engineering on the design of a new dustpan, ". . . and it's got to be beautiful," says the boss.

The value man and the designer look at each other.

"A *beautiful* dustpan? . . . !" asks the designer.

At this point the value specialist has discovered one of his electives: *Industrial Design.* Then the Small Business Administration helps the little company get a government contract, with a value engineering clause in it. Another elective: *Government Contract Management.* Here is a sample of electives in the value disciplines:

Table C-3   Technical Electives

Course	Credits
Industrial Design	3
Business Law	2
Critical Path Networks	2
Product Safety	2
Quality Assurance	3
Reliability	2
Marketing Research	2
The Computer in Business	2
Government Contract Management	2

From these 20 credit hours in technical electives the student would not be allowed to choose more than 10, leaving eight for liberal studies and six to personal choice, outside his major.

We have easily enough college-level instructional material for a four-year course in value improvement, but it is later than we think . . . .

DEGREE COURSES

Suppose your boss said, "You have to recruit value specialists while they are still in college. I want people who can stand toe-to-toe with ME's, IE's, and MBA's."

"All my people have degrees," you answer, "and some have advanced—"

"No," he cuts you off. "I don't want ME's or IE's or MBA's who have been varnished over with value in a two-week seminar. I want graduate value analysts or value engineers or whatever the college wants to call them."

You could ask, "What college offers degrees in the value disciplines?" But bosses usually want answers, not other questions.

We have seen, so far, that the professional foundations of the value disciplines amply justify organizing them into degree courses. One problem, however, is that our value improvement work, up to this point, has not been clearly identified with its professional foundations. Another problem is that some of the names we have for the value specialist are ambiguous and—from an academic point of view—misleading.

Here is a typical conversation with a dean of academic affairs.

"What is the difference," he asks, "between value engineering and engineering as it is taught here, which includes engineering economics and cost analysis?"

You tell him.

"Ah," he acknowledges, showing interest, "there is a big difference. What an intriguing combination of physical and social sciences. No wonder industry demands it. Throughout the university we offer many suitable courses—business economics, cost control methods, industrial psychology, creative problem solving. All this points up to a most useful major in industrial science. You will have to find a suitable professional title—"

"What's wrong with value engineering?"

He looks at you in surprise. "You just showed me how much broader it is than engineering. It looks like a very worthwhile profession in its own right, but engineering it is not."

"I am an engineer," you point out.

He smiles compassionately. "I know, I am an experimental psychologist, and here I am in administration. It happens to the best of us."

"But I am happy in my work!"

"So am I, but that doesn't mean that experimental psychology is administration or that whatever you do to value is engineering—what *do* you do to value?"

"Analyze it!"

"You analyze value . . ." he ponders; then, beaming, "What's wrong with value analysis?"

It turns out that of all our labels *value analysis* is the most acceptable to the universities. I am sure that other terms are—it depends on the school—but *value analysis* is the name of my own operation, and I can offer it to a chief engineer without having him say anything about a 10-foot pole.

AN ACADEMIC END-RUN

Speaking about 10-foot poles, the more an academic discipline is well-established and time-honored, the more 10-foot poles it has with which not to touch anything new. On the other hand, the recent invaders of the halls of learning are too busy getting organized to worry about 10-foot poles. Entering the dean's office in a vigorous, modern institution, you say, "Dean Roe? . . . I am planning a course on value analysis and—"

"Grab the other end of this table," he interrupts you. "It belongs in Industrial Engineering."

"Value analysis does?"

"No, the table. We're moving to a fine new industrial engineering building."

In the hallway he lowers his end of the table. "We can leave it here till the men come for it." Sizing you up for the first time, he asks, "Philosopher chap? . . . you mentioned value—"

"No, I am in industry."

"In industry? Heck, we can carry the table ourselves. On to the new building!"

As you stumble across the campus, he says over his shoulder, "Value, in industry, *does* belong in Industrial Engineering. It's our cup of tea."

A well-planned course on value analysis is also welcome in the School of Business Administration or the School of Management.

Where does it belong? Ah . . . that's the problem. It belongs in *both!* Many universities are providing the means for combining human knowledge into useful blends unheard of yesterday. As one professor puts it, "We cannot let the buildings tell us what we can teach and where."

In industry value analysis has developed a bridge between the physical sciences of engineering and manufacturing, on the one hand, and the socioeconomic sciences of finance, marketing, and purchasing, on the other. The university, having a wealth of such bridges, may consent to teach people how to use *our* bridge; might even let us set it up between Industrial Engineering and Business Administration. Here is the proposal:

MASTER OF SCIENCE IN VALUE ANALYSIS

Requirements for such a degree as they would appear in the Graduate School catalog:

Admission Requirements

1. A bachelor's degree from an approved institution.

2. An undergraduate major in engineering, physics, metallurgy, or a field considered suitable by the admissions committee.

3. Seventy credit hours of academic work outside the undergraduate major.

4. An undergradute grade-point average of at least 2.15 on a basis of 4.

Subject to the judgment of the department chairman, professional registration and successful practice in engineering may fulfill the admission requirements.

Graduate Work. Thirty credits of graduate work are required, at least nine of which must be in value analysis subjects, nine in industrial engineering subjects related to value analysis, six in Business Administration, and six in Psychology for Value Analysts. A thesis is optional. If a thesis is presented, the total number of credits required is only 24, at least eight of which must be in industrial engineering subjects.

MASTER OF ARTS IN VALUE ANALYSIS

Here are the requirements for such a degree as they would appear in the catalog of a Graduate School of Business.

Admission Requirements

1. A bachelor's degree from an approved institution.

2. Completion of at least 55 credit hours in science and engineering, as well as 35 credit hours in the social sciences.

3. Passing the Admissions Test for Graduate Study in Business.

Graduate Study. Thirty credits of graduate work are required, at least nine of which must be in value analysis subjects, nine in business administration subjects related to value analysis, six in Science and Engineering, and six in Psychology for Value Analysts. A thesis is optional. If a thesis is presented, the total number of credits required is only 24, at least eight of which must be in Business Administration.

Table C-4   IE Subjects Related to VA
(37 credit hours available)

Subject	Credits
Economic Decision Making	2
Theory of Design	2
Human Factors Engineering	3
Mechanics of Materials	2
Manufacturing Processes	3
Work Design	3
Elementary Statistical Analysis	3
General Economics	4
Engineering Economy and Analysis	3
Introduction to Materials Science	3
Introduction to Optimization Methods	3
Linear Programming	3
Dynamic Programming	3

Table C-5   Business Subjects Related to VA
(45 credit hours available)

Subject	Credits
Organization: Individual and Group Behavior	2
Product Innovation and Marketing	3
Marketing Cases and Problems	3
Business Statistics	3
Introduction to Decision Theory	3
Economics of the Consumer and the Firm	3
Cost Accounting	2
Cost Administration and Control	2
Budgets and Budgeting Control	2
Industrial Management	3
Production Planning and Control	3
Industrial Purchasing	2
Marketing	3
Marketing Cases and Problems	3
Micro-Organization Behavior	3
Mathematical Analysis for Business	3
Operational Decision in Production Management	2

Table C-6 Psychology for Value Analysis
(15 credit hours available)

Subject	Credits
Psychology of Motivation	3
Thinking and Problem Solving	3
Introductory Social Psychology	3
Social Behavior Dynamics	3
Group Dynamics	3

WHAT TO DO NOW

With the wealth of course material available among value specialists and in the universities it is relatively simple to design undergraduate courses in value analysis. One such course could lead to a BS in Value Analysis or Value Technology, two others could be BSIE and BSBA, each with Value Analysis option.

But, like a product, a course must sell. The value disciplines must be better known by the college faculty and student body before youngsters can even consider going into value analysis. One way of conveying information on the value disciplines to college faculty members is in the University Extension Division.

A UNIVERSITY EXTENSION COURSE

Rutgers, The State University   **Camden Campus**

Value Analysis
14 Wednesdays 6:15 to 7:55 p.m.

I. Introduction   *Session 1*
   The concept of value
   The principle of consumer sovereignty
   Some working definitions
   Origin and development of value analysis

II. Organizing a Value Workshop   *Session 2*
    Composition of the teams
    Riddle of group dynamics
    The value task group

REFERENCES

Ackoff, Russell L., 1962. *Scientific Method*, Wiley, New York.

Alamshah, William H., 1967. "The Conditions for Creativity," *J. Creative Behavior*, 1, No. 3 (Summer).

American Society of Tool and Manufacturing Engineers, 1967. *Value Engineering in Manufacturing*, Prentice-Hall, Englewood Cliffs, N. J.

Beardsley, Monroe C., 1950. *Practical Logic*, Prentice-Hall, Englewood Cliffs, N. J.

Benson, Bernard S., 1957. "Let's Toss this Idea Up," *Fortune* (October).

Beveridge, Gordon S. G., and Robert S. Schechter, 1970. *Optimization: Theory and Practice*, McGraw-Hill, New York.

Born, Max, 1964. *Natural Philosophy of Cause and Chance*, Dover, New York.

Bower, Marvin, 1966. *The Will to Manage*, McGraw-Hill, New York.

Boyer, Carl B., 1968. *A History of Mathematics*, Wiley, New York.

Brockmeyer, E., H. L. Halstrom, and Arne Jensen, 1948. The Life and Works of A. K. Erlang, *Trans. Danish Acad. Tech. Science*, No. 2, Copenhagen.

Bross, Irwin D. J., 1953. *Design for Decision*, Macmillan, New York.

Burington, R. S., and D. C. May, 1958. *Handbook of Probability and Statistics with Tables*, Handbook Publishers, Sandusky, Ohio.

Campbell, Norman Robert, 1957. *Physics: the Elements*, Cambridge University Press; also Dover, New York.

Campbell, Norman Robert, 1921. *What Is Science*, Dover, New York.

Carnap, R., and Y. Bar-Hillel, 1953. "An Outline of a Theory of Semantic Information," M.I.T. Research Lab. Electronics, Tech. Rept. 247.

Chernoff, H., and L. E. Moses, 1959. *Elementary Decision Theory*, Wiley, New York.

Cherry, Colin, 1957. *On Human Communication,* M.I.T. Press, Cambridge, Mass.

Churchman, C. W., 1961. *Prediction and Optimal Decision,* Prentice-Hall, Englewood Cliffs, N. J.

——, 1959. "Why Measure?" in C. W. Churchman, and Philburn Ratoosh, Eds., *Measurement Definitions and Theories,* Wiley, New York.

Churchman, C. W., R. L. Ackoff, and E. L. Arnoff, 1957. *Introduction to Operations Research,* Wiley, New York.

——, and Philburn Ratoosh, Eds., 1959. *Measurement Definitions and Theories,* Wiley, New York.

Clarke, Grahame, and Stuart Piggot, 1965. *Prehistoric Societies,* Knopf, New York.

Coombs, C. H., 1952. "A Theory of Psychological Scaling," *Bull. Eng. Res.,* **34,** University of Michigan Press, Ann Arbor.

Cotlow, Lewis, 1966. *In Search of the Primitive,* Little, Brown, Boston.

Cox, D. R., and Walter L. Smith, 1961. *Queues,* Methuen Monographs, Wiley, New York.

Department of Defense, 1968. *DoD Handbook 5010.8-H,* Superintendent of Documents, U. S. Government Printing Office, Washington, D. C. 20402.

Drucker, Peter F., 1967. *The Effective Executive,* Harper & Row, New York.

Dulles, Allan, 1963. *The Craft of Intelligence,* Harper & Row, New York.

Ellis, Brian, 1966. *Basic Concepts of Measurement,* Cambridge University Press, London and New York.

Falcon, William D., Ed., 1964. *Value Analysis Value Engineering,* American Management Association, New York.

Farb, Peter, 1968. *Man's Rise to Civilization,* Dutton, New York.

Fayol, Henri, 1949. *General and Industrial Administration,* English translation by Constance Storrs of *Administration Industrielle et Générale,* Dunod, Paris (1925), Pitman, London.

Feller, William, 1966. *An Introduction to Probability Theory and Its Applications,* Volume 2, Wiley, New York.

——, 1958. *An Introduction to Probability Theory and Its Applications,* Volume 1, Wiley, New York.

Ferguson, Robert O., and Lauren F. Sargent, 1958. *Linear Programming: Fundamentals and Applications,* McGraw-Hill, New York.

Fishburn, Peter C., 1964. *Decision and Value Theory,* Wiley, New York.

Follett, Mary Parker, 1924. *Creative Experience,* Longmans, Green, New York.

Frondizi, Risieri, 1963. *What Is Value,* Open Court Publishing Co., La Salle, Ill.

Fry, Thornton C., 1965. *Probability and Its Engineering Uses,* Van Nostrand, Princeton, N. J.

Gabor, D., 1951. *Lectures on Communication Theory,* M. I. T., Cambridge, Mass.

Galton, F., 1869. *Hereditary Genius,* Appleton, New York.

Gantt, Henry, 1916. *Industrial Leadership,* Associated Press, New York.

Garvin, Walter, 1960. *Introduction to Linear Programming,* McGraw-Hill, New York.

Gilbert, S. J., 1967. "A study of some influences on the creativeness of da Vinci and Michelangelo, during their early years, by the culture and society of that period," *J. Creative Behavior,* **1,** No. 2.

Gilbreth, Frank B., and Lillian, E. M. Gilbreth, 1912. *Primer of Scientific Management,* Van Nostrand, New York.

Gilbreth, Lillian, and Alice Rice Cook, 1947. *The Foreman and Manpower Management,* McGraw-Hill, New York.

Gordon, William J. J., 1961. *Synectics,* Harper & Row, New York.

Grayson, C. J., Jr., 1960. *Decisions Under Uncertainty,* Harvard, Cambridge, Mass.

Griffith, Samuel B., 1963. *Sun Tzu, The Art of War,* Oxford University Press, London and New York.

Guilford, J. P., 1967. *The Nature of Human Intelligence,* McGraw-Hill, New York.

Hadamard, Jaques, 1954. *The Psychology of Invention in the Mathematical Field,* Princeton University Press, Princeton, N. J., and Dover, New York.

Hall, Arthur D., 1962. *A Methodology for Systems Engineering,* Van Nostrand, Princeton, N. J.

Hall, Everett W., 1961. *Our Knowledge of Fact and Value,* University of North Carolina Press, Chapel Hill, N. C.

Hammersley, J. M., and D. C. Handscomb, 1964. *Monte Carlo Methods,* Methuen Monographs, Wiley, New York.

Hare, Van Court, Jr., 1967. *Systems Analysis: A Diagnostic Approach,* Harcourt, Brace, & World, New York.

Hartley, R. V. L., 1928. "Transmission of Information," *Bell System Tech. J.,* 7, (July).

Heller, Edward D., 1971, *Value Management, Value Engineering, and Cost Reduction,* Addison-Wesley, Reading, Mass.

Hess, S. W., and H. A. Quickley, 1963. "Analysis of Risk in Investments Using Monte Carlo Techniques," *Chem. Eng. Progr. Symp. Ser.* 42, 59:55.

Heyel, Carl, Ed., 1963. *The Encyclopedia of Management,* Reinhold, New York.

Hillier, Frederick S., and Gerald J. Lieberman, 1967. *Introduction to Operations Research,* Holden-Day, San Francisco.

Hodgman, C. E., 1959. *CRC Standard Mathematical Tables,* Chemical Rubber Publishing Co., Cleveland, Ohio.

Hoel, Paul G., 1954. *Introduction to Mathematical Statistics,* Wiley, New York.

Hohn, Franz E., 1964. *Elementary Matrix Algebra,* Macmillan, New York.

Householder, A. S., G. E. Forsythe, and H. H. Germond, Eds., 1951. *Monte Carlo Method,* National Bureau of Standards, U. S. Government Printing Office, Washington, D. C.

Jouineau, Claude, 1968. *L'Analyse de la valeur,* Enterprise Moderne d'Edition, 4, rue Cambon, Paris (1er).

Juran, J. M., Leonard A. Seder, and Frank M. Gryna, Jr., 1962. *Quality Control Handbook,* McGraw-Hill, New York.

Kappel, Frederick R., 1960. *Vitality in a Business Enterprise,* McGraw-Hill, New York.

Karlins, M., R. Lee, and H. Schroder, 1967. "Creativity and Information Search in a Problem Solving Context," *Psychol. Sci.,* 8.

Kemeny, J. G., A. Schleifer, Jr., J. L. Snell, and G. L. Thompson, 1962. *Finite Mathematics with Business Applications,* Prentice-Hall, Englewood Cliffs, N. J.

Kepner, C. H., and B. B. Tregoe, 1965. *The Rational Manager*, McGraw-Hill, New York.

Khintchine, A. Y., 1960. *Mathematical Methods in the Theory of Queueing*, Charles Griffin, London, and Hafner, New York.

Kline, Morris, 1959. *Mathematics and the Physical World*, Crowell, New York.

Koestler, Arthur, 1964. *The Act of Creation*, Macmillan, New York.

Kolmogoroff, A. N., 1933. *Warscheinlichkeitsrechnung*, Springer-Verlag, Berlin. See following entry for translation.

——— (1950 English translation of the above, edited by Nathan Morrison). *Foundations of the Theory of Probability*, Chelsea, New York.

Kourim, Gunther, 1968. *Wertanalyse*, R. Oldenbourg, Munich and Vienna.

Kuhn, Thomas S., 1963. "The Essential Tension," in C. W. Taylor and F. Barron, Eds., *Scientific Creativity*, Wiley, New York.

Kuo Hua-Jo, 1957. *A Modern Translation of Sun Tzu's Art of War* (from ancient into modern Chinese). Peoples Press, Peking.

Kuo Hua-Jo, 1939. *A Preliminary Study of Sun Tzu's Art of War* (in Chinese), Fushih, Shensi Province, China.

Lamont, W. D., 1955. *The Value Judgement*, Edinburgh University Press, Edinburgh, Scotland. Out of print but available in university libraries.

Lanczos, Cornelius, 1965. *Albert Einstein and the Cosmic World Order*, Wiley: Interscience, New York.

Leroi-Gourhan, André (in print in 1969, no copyright notice). *Treasures of Prehistoric Art*, Harry N. Abrams, New York.

Levin, Richard I., and C. A. Kirkpatrick, 1965. *Quantitative Approaches to Management*, McGraw-Hill, New York.

Levy, Lester S., and Roy J. Sampson, 1962. *American Economic Development*, Allyn & Bacon, Boston.

Lewin, Kurt, 1951. *"Field Theory in the Social Sciences,"* Dorwin Cartwright, Ed., Harper, New York.

Loomba, N. Paul, 1964. *Linear Programming: an Introductory Analysis*, McGraw-Hill, New York.

Luce, Duncan R., and Howard Raiffa, 1958. *Games and Decisions*, Wiley, New York.

MacDuffee, Cyrus Colton, 1943, but very good! *Vectors and Matrices*, Carus Monograph No. 7, Mathematical Association of America.

McGraw-Hill Dictionary of Economics, 1966. McGraw-Hill, New York.

Machol, Robert E., and Paul Gray, 1962. *Recent Developments in Information and Decision Processes*, Macmillan, New York.

McKinsey, J. C. C., 1952. *Introduction to the Theory of Games*, McGraw-Hill, New York.

Majumdar, Tapas, 1961. The Measurement of Utility, MacMillan, London and New York.

Marcus, Marvin, and Henryk Minc, 1965. *Introduction to Linear Algebra*, Macmillan, New York.

Marshall, Alfred, 1966 ed. *Principles of Political Economy*, Macmillan, London (first published in 1890).

Metcalf, H. C., and L. Urwick, 1941. *Dynamic Administration: The Collected Papers of Mary Parker Follett*, Pitman, London.

Metropolis, N., and S. Ulam, 1949. "The Monte Carlo Method," *J. Am. Statist. Assoc.*, 44, No. 247, 335–341.

Meyer, Herbert A., Ed., 1956. *Symposium on Monte Carlo Methods*, Wiley, New York.

Meyers, Albert L., 1942. *Elements of Economics*, Prentice-Hall, Englewood Cliffs, New Jersey.

Miles, Lawrence D., 1968. "The Challenges that Lie Ahead," *Proceedings of the 10th Anniversary Conference, Value Analysis Educational Program*, Value Analysis, Schenectady, N. Y.

———, 1966. "The Fundamentals of Value Engineering, *Proceedings of the 1966 National Convention*, American Society of Value Engineers, Smyrna, Ga.

———, 1961. *Techniques of Value Engineering and Analysis*, McGraw-Hill, New York. Rev. ed., 1971.

Mill, John Stuart, 1965 ed. *Principles of Political Economy*, Augustus M. Kelley, New York.

Molina, E. C., 1942. *Poisson's Exponential Binomial Limit*, Van Nostrand, New York.

Moore, A. D., 1969. *Invention, Discovery, and Creativity*, Doubleday, Garden City, N. Y.

Mudge, Arthur E., 1971, *Value Engineering*, McGraw-Hill, New York.

National Academy of Sciences, 1969. *Value Engineering in Federal Construction Agencies SBN-309-01756-4*, NAS Printing Office, 2101 Constitution Ave., Washington, D. C. 20418.

Newman, James R., 1956. *The World of Mathematics*, Simon & Schuster, New York. Vol. 4, p. 2041; also a biographical sketch of Poincaré in Vol. 2, p. 1374.

Nyquist, Harry, 1917, "Certain Factors Affecting Telegraph Speed," *Bell System Tech. J. 3*, 1924.

Odiorne, George S., 1965. *Management by Objectives*, Pitman, New York.

Ohlin, Bertil, 1967. *Interregional and International Trade*, Harvard University Press, Cambridge, Mass.

Ollner, Jan, et al., 1967. *Vardeanalys*, Sveriges Mekanforbung, Stockholm.

Osborn, Alex F., 1961. *Applied Imagination*, Scribner's, New York.

Pareto, Vilfredo, 1917. *Traité de sociologie générale*, Paris; translated as *Mind and Society*, Dover, New York, 1963.

Pareto, Vilfredo, 1897. *Cours d'économie politique*, Lausanne. Includes "Pareto's law" of uneven distribution.

Parzen, Emanuel, 1962. *Stochastic Processes*, Holden-Day, San Francisco.

———, 1960. *Modern Probability Theory and Its Applications*, Wiley, New York.

Pascal, Blaise, 1670. *Pensées*. IV-227, Paris.

Patrick, Catherine, 1937. "Creative Thought in Artists," *J. Psychol.*, 26, 35–73.

Pierce, J. R., 1961. *Symbols, Signals and Noise*, Harper & Row, New York.

Poincaré, Henri (in print 1971). *Science and Method*. Dover, New York.

Purchasing Magazine, 1968. *Value Analysis Issue*, New York, May 16.

RAND Corporation, 1955. *A Million Random Digits*, Free Press, Glencoe, Ill.

Rathe, Alex W., Ed., 1961. *Gantt on Management*, American Management Association, New York.

Reichenbach, Hans, 1957. *Space and Time*, Dover, New York.

Ridge, Warren J., 1969. *Value Analysis for Better Management*, American Management Association, New York.

Riordan, John, 1962. *Stochastic Service Systems*, Wiley, New York.

Saaty, Thomas L., 1961. *Elements of Queueing Theory*, McGraw-Hill, New York.

Sandars, N. K., 1960. *The Epic of Gilgamesh*, Penguin, Baltimore.

Sasieni, Maurice, Arthur Yaspan, and Lawrence Friedman, 1959. *Operations Research: Methods and Models*, Wiley, New York.

Savage, L. J., 1954. *The Foundations of Statistics*, Wiley, New York.

Savage, L. J., et al., 1962. *The Foundations of Statistical Inference*, Wiley, New York; Methuen, London.

Schiller, F. C. S., 1912. *Formal Logic*, Macmillan, London.

Schlaifer, Robert, 1961. *Introduction to Statistics for Business Decisions*, McGraw-Hill, New York.

Schumpeter, J. A., 1934. *The Theory of Economic Development*, Harvard University Press (translated from the 1911 German edition), Cambridge, Mass.

Shannon, Claude E., 1948. "A Mathematical Theory of Communication," *Bell System Tech. J.*, **27**, (July-October).

Shannon, Claude E., and Warren Weaver, 1949. *The Mathematical Theory of Communication*, University of Illinois Press, Urbana.

Shreider, Yu. A., Ed., 1966. *The Monte Carlo Method*, Pergamon, New York.

Smart, William, 1966 ed. *Introduction to the Theory of Value*, Augustus M. Kelley, New York.

Snedecor, George W., 1956. *Statistical Methods*, Iowa University Press, Ames, Iowa.

Somer, Louise, 1954. "Exposition of a New Theory on the Measurement of Risk," a translation of Daniel Bernoulli's (1738) "Specimen Theorial Novae de Mensura Sortis," *Econometrica*, **22**.

Sténuit, Robert, 1969. "Priceless Relics of the Spanish Armada," *National Geographic*, Washington, June issue.

Stevens, S. S., 1959. "Measurement, Psychophysics, and Utility" in *Measurement Definitions and Theories*, C. W. Churchman and Philburn Rotoosh, Eds., Wiley, New York.

Stevens, S. S., 1946. "On the Theory of Scales of Measurement," *Science*, **103**, 677–680.

Stewart, Frank M., 1963. *Introduction to Linear Algebra*, Van Nostrand, Princeton, N. J.

Taylor, Frederick W., 1911. *The Principles of Scientific Management*, Harper, New York.

Thrall, Robert M., and Leonard Tornheim, 1957. *Vector Spaces and Matrices*, Wiley, New York.

U. S. Department of Defense, January 1967. *Reduce Cost and Improve Equipment Through Value Engineering*, U. S. Government Printing Office, Washington, D. C.

Value Engineering, Bruce D. Whitwell, Ed. Published bimonthly by Pergamon Press, Oxford, England, and 4401-21st St., Long Island City, New York 11101.

Value Engineering, Journal of, W. B. Dean, Ed., Published quarterly by the Society of American Value Engineers, 410 West Verona St., Kissimme, Florida 32741.

Value Engineering Digest, Richard J. O'Connell, publisher. Issued twice a month by *Sci/Tech Digests*, Washington, D. C. 20004 (a must for the professional).

Van Doren Stern, Philip, 1969. *Prehistoric Europe*, Norton, New York.

Vasonyi, Andrew, 1958. *Scientific Programming in Business and Industry*, Wiley, New York.

Veblen, Thorstein, 1967. *The Theory of the Leisure Class*, Funk & Wagnalls, New York.

Von Neumann, John, and Oskar Morgenstern, 1953. *Theory of Games and Economic Behavior*, Princeton University Press, Princeton, New Jersey.

Wallas, G., 1945. *The Art of Thought*, Watts, London, 1926 and 1945.

Wallis, Allen W., and Harry V. Roberts, 1956. *Statistics, A New Approach*, Free Press, Glencoe, Ill.

Watanabe, Salosi, 1969. *Knowing and Guessing*, Wiley, New York.

Weisselberg, Robert C., and Joseph G. Cowley, 1969. *The Executive Strategist*, McGraw-Hill, New York.

Whittaker, Edmund, 1960. *Schools and Streams of Economic Thought*, Rand McNally, Chicago, and John Murray, London.

Wiener, Norbert, 1948. *Cybernetics*, Wiley, New York.

Williams, J. D., 1954. *The Compleat Strategist*, McGraw-Hill, New York.

Wilson, Ira G., and Marthann E. Wilson, 1965. *Information, Computers, and System Design*, Wiley, New York.

Xenophon, 1968. *Cyropaedia*, Loeb Classical Library, Harvard, Cambridge, Mass.

INDEX